thods f
dings.

Predictive Methods for the Energy Conserving Design of Buildings

Predictive Methods for the Energy Conserving Design of Buildings

edited by

Henry J. Cowan
Professor of Architectural Science
University of Sydney, Australia

Papers presented as part of the Architecture Section of the 52nd Conference of the Australian and New Zealand Association for the Advancement of Science at Macquarie University, Sydney, 12 May 1982

PERGAMON PRESS
SYDNEY • OXFORD • NEW YORK • TORONTO • PARIS • FRANKFURT

Pergamon Press (Australia) Pty Ltd,
19a Boundary Street, Rushcutters Bay, N.S.W. 2011, Australia.

Pergamon Press Ltd,
Headington Hill Hall, Oxford OX3 0BW, England.

Pergamon Press Inc.,
Maxwell House, Fairview Park, Elmsford, N.Y. 10523, U.S.A.

Pergamon Press Canada Ltd,
Suite 104, 150 Consumers Road, Willowdale, Ontario M2J 1P9, Canada.

Pergamon Press GmbH,
6242 Kronberg-Taunus, Hammerweg 6, Postfach 1305, Federal Republic of Germany.

Pergamon Press SARL,
24 rue des Ecoles, 75240 Paris, Cedex 05, France.

First published 1983

Cover design by Robert Taylor
Printed in Hong Kong by Warren Printing Co. Ltd

National Library of Australia Cataloguing in Publication Data:

Predictive methods for energy conserving design
of buildings.

Symposium held at Macquarie University, Sydney
on 12 May 1982 as part of the programme of the
Architecture Section of the 52nd ANZAAS Congress.
Includes index.

ISBN 0 08 029838 9.

1. Buildings — Thermal properties — Data
processing — Congresses. I. Cowan, H. J.
(Henry Jacob), 1919- . II. ANZAAS. Congress

Contents

Preface

Since 1974 a great deal of theoretical work has been undertaken on the prediction of the thermal performance of buildings, and many computer programs of varying precision have been written for this purpose. Some are so complex that the increased cost of design does not compensate for the energy savings achieved in the small buildings that comprise the practice of most architects. The simpler programs, on the other hand, do not necessarily model the thermal performance with the requisite accuracy. There is, moreover, a lack of experimental data on the thermal performance of buildings.

A symposium on "Predictive Methods for the Energy Conserving Design of Buildings" was held to examine these problems at Macquarie University, Sydney, on 12 May 1982, as part of the programme of the Architecture Section of the 52nd Conference of the Australian and New Zealand Association for the Advancement of Science. This volume contains an edited version of the papers presented at that symposium.

The first two papers describe two simple computer programs developed by Mr Wooldridge at the CSIRO Division of Mechanical Engineering (now the Division of Energy Technology) and by Dr Szokolay at the Architectural Science Unit of Queensland University.

In the third paper Mr Williamson compares the predictions of four different computer programs, which are not identified, with the measured thermal performance of two buildings in Melbourne and in Brisbane.

Mr Charters describes an experimental investigation of the prototype for the Merchant Builders Trombe Wall House in Melbourne, and compares the experimental thermocirculation heat transfer with the theoretical prediction.

Dr Carruthers compared thermal measurements undertaken on houses in Western Australia with the results of calculations made with the aid of two computer programs, developed by the University of Wisconsin and the CSIRO Division of Building Research.

The final paper by Mr Smart and Mr Ballinger gives the detailed measurements undertaken on twelve houses on a new estate, constructed by the New South Wales Housing Commission, and an analysis of the results obtained.

This volume makes these important and practically useful papers generally available.

H.J.C.
Department of Architectural Science,
University of Sydney,
Australia

August 1982

List of Contributors

MR JOHN A. BALLINGER
Senior Lecturer in Architecture, Director of Solarch,
University of New South Wales,
Kensington, New South Wales 2033.

DR DEREK D. CARRUTHERS
Senior Lecturer in Architecture,
University of Western Australia, Nedlands,
Western Australia 6009.

MR WILLIAM W. S. CHARTERS
Reader in Thermodynamics, University of Melbourne,
and Chairman of the Victorian Solar Energy Council,
270 Flinders Street, Melbourne, Victoria 3000.

MR MICHAEL SMART
Professional Officer, Solarch, Solar Architecture
Research Unit, University of New South Wales,
Kensington, New South Wales 2033

DR STEPHEN V. SZOKOLAY
Reader in Architecture, University of Queensland,
St Lucia, Queensland 4067

MR TERRY WILLIAMSON
Lecturer in Architecture, University of Adelaide,
Adelaide, South Australia 5000

MR MICHAEL J. WOOLDRIDGE
Principal Research Scientist, Division of Energy
Technology, Commonwealth Scientific and Industrial
Research Organization, Highett, Victoria 3190

A New Mini-Computer Program to Predict Building Energy Use

M.J. Wooldridge
CSIRO Division of Energy Technology,
Highett, Victoria

ABSTRACT

This paper describes work in progress in a project, partly funded by the National Energy Research, Development and Demonstration Council, being conducted by the Divisions of Building Research and Energy Technology of CSIRO. In particular it concentrates on the package for the computation of energy consumption in non-residential buildings being created at the Division of Energy Technology.

The project objectives are:

> To produce a set of computer-based design procedures and guidelines for the thermal design of all types of buildings in conjunction with their associated energy consuming systems. Specifically, the creation of a set of routines adequately describing the performance of most commonly used (Australian) buildings and plant; and the combination of these into a suite of programs capable of assessing energy consumption over a specific period of time for different building types. Separately, to produce a climatic data bank for several Australian locations, suitable for input to such programs.

The concept of a set of design and analysis procedures for such buildings and plant is seen as providing a nationally acceptable basis for assessing building energy consumption. Considerable emphasis in the project is being given to user requirements through the involvement in an advisory capacity of a group of practising engineers from both private industry and government.

The project is due to be completed in late 1982.

The work programme comprises the following steps:

- identification of existing methodology and programs in the field,
- creation of methodology suitable for Australian building needs, if necessary,
- identification of heating, cooling and air distribution systems to be modelled, and how they are to be combined into a number of programs,
- production of a flow chart and schedule for each program ensuring that 'external' comment is made on the proposed software,

- creation of the routines describing the thermal performance of each of the systems identified as candidates for inclusion in the package,
- integration of the above routines into the computer programs of the suite,
- testing of the programs and incorporation of user feedback,
- assembly of building and climatic data bank.

Having surveyed existing software in this field, mainly that available from US sources, the project team determined on the following approach:

- develop and package a small computer procedure for computing energy consumption in buildings, which would meet the perceived Australian need,
- obtain DOE-2 and mount it on the CSIRONET computing network.

The remainder of this paper describes the package being developed, christened BUNYIP (Building Energy Investigation Package), to meet the Australian need for a design procedure which can be used in the early analysis of a building project when firm decisions are usually taken regarding choice of energy consuming equipment.

INTRODUCTION

This paper is a progress report on a project, nearing completion, one of the aims of which is to produce a computing package for regular use in Australian design offices for the estimation of building energy use. After summarizing the building energy situation in Australia, the paper discusses the project approach. The characteristics of the package being written are described, but its detail is the subject of special support documentation presently being prepared.

ENERGY USE IN BUILDINGS

The secondary energy dissection in Australia, with each major usage, is shown on Fig. 1 (Ref. 1). Buildings account for some twenty per cent of the secondary energy consumption: almost half of this is electricity; gas and oil each account for approximately twenty per cent. Of the building sector consumption, residential usage accounts for some two thirds and commercial, institutional and other non-residential buildings account for the remainder. In general gas and electricity usage is increasing faster than in the two other sectors (industry and transport), while oil use is generally declining in the building sector. The total market value of energy used in buildings is in the order of $2,000 million a year. Table 1 (Ref. 2) shows typical usage in a few commercial buildings in Australia, illustrating that building energy consumption per year is more a function of the usage of the building, rather than of the climate. Office buildings that are occupied ten hours per day range in energy use between 0.6 and 2.0 GJ per square metre per year, while twenty-four hour occupancy buildings in the commercial/institutional sector generally ranged between 1.5 and 3.5 GJ per square metre per year.

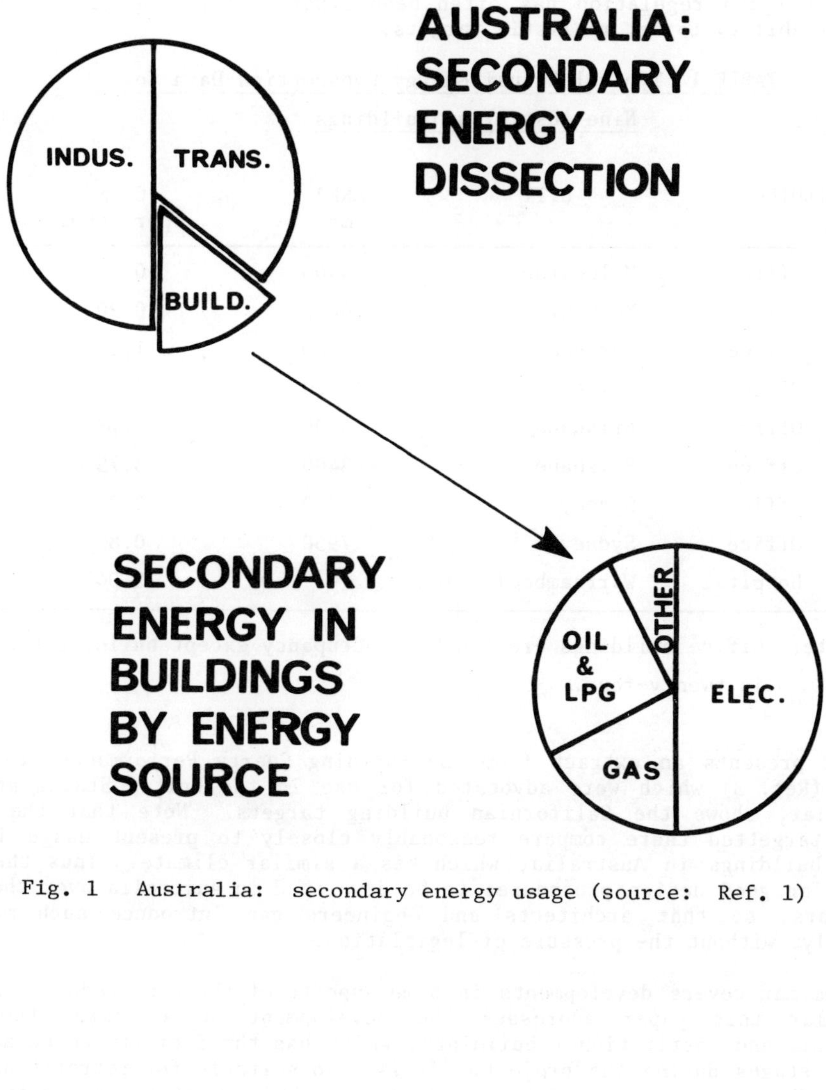

Fig. 1 Australia: secondary energy usage (source: Ref. 1)

The question may then be asked, "Is regulation necessary to control the rate of energy usage in Australian buildings?". The present author suggests that a voluntary process based on economic factors and commonsense would enable existing design methods, or design methods presently being produced, to be evaluated before introducing regulations controlling building energy use. This policy would give time for the design procedures to be evaluated, and the setting of target consumption figures which were both reasonable and attainable. The experience overseas of an attempted rapid introduction of legislation and regulation has often been frustrated by the lack of design tools or ability to set realistic targets.

TABLE 1 Typical Annual Energy Consumption Data for Nine Australian Buildings

BUILDING		CITY	AREA m^2	GJ/m^2 per annum
1.	Office	Melbourne	4400	0.56
2.	Office	Melbourne	16000	0.90
3.	Office	Horsham, Vic	2000	1.17
4.	Office	Brisbane	15000	0.70
5.	Office	Brisbane	14300	0.73
6.	Office	Brisbane	3400	3.75
7.	Office	Sydney	7900	0.85
8.	Office	Sydney	7950	0.83
9.	Hospital	Warrnambool, Vic		2.04

Note: Office buildings are ten hour occupancy except building 6 which is twenty-four.

Table 2 presents an extract from the Building Energy Performance Standards (BEPS) (Ref. 3) which were advocated for use in the United States and, in particular, shows the Californian building targets. Note that the usage levels targetted there compare reasonably closely to present usage in the better buildings in Australia, which has a similar climate. Thus the need exists for good design procedures to be developed in Australia over the next few years, so that architects and engineers can introduce such methods gradually, without the pressure of legislation.

This seminar covers developments in some aspects of those procedures, and in particular this paper addresses the development of a design tool for commercial and institutional buildings, which has the facility to be used at several stages during the project. It is also suitable for retrofit assessments. The package has been produced at the Division of Energy Technology, CSIRO, with support from the National Energy Research, Development and Demonstration Council over the period 1980/82. It has been christened BUNYIP (Building Energy Investigation Package) and is suitable for use on micro to large main frame computers. The input sheets are simple to use and the model

TABLE 2 Extract from US Government Federal Register (Ref. 3)

Federal Register / Vol. 44, No. 230 / Wednesday, November 28, 1979 / Proposed Rules

ENERGY BUDGET LEVELS FOR COMMERCIAL AND MULTIFAMILY RESIDENTIAL BUILDING DESIGNS
(in MBtu/sq. ft./yr)

State	SMSA	Clinic	Community Center	Gymnasium	Hospital	Hotel Motel	Multifamily High Rise	Multifamily Low Rise	Nursing Home	Office Large	Office Small	School Elementary	School Secondary	Shopping Center	Store	Theater/Auditorium	Warehouse
Alabama	Birmingham	123	107	127	353	166	114	110	161	113	101	89	117	181	142	139	53
	Mobile	142	129	147	406	192	127	132	187	131	116	96	133	207	166	162	47
Arizona	Phoenix	146	133	152	406	196	131	136	192	134	119	100	137	212	171	168	49
California	Bakersfield	123	109	127	358	167	113	112	162	113	100	86	116	181	143	140	48
	Fresno	120	105	123	353	163	112	108	158	111	98	85	114	178	139	136	50
	Los Angeles	112	101	115	364	157	103	103	151	106	91	74	106	171	132	126	42
	Oakland	108	93	108	353	150	102	94	143	101	87	75	103	164	125	119	50
	Sacramento	118	102	120	353	160	110	104	154	108	96	84	112	175	136	132	52
	San Diego	114	103	117	364	158	104	106	153	107	92	75	107	172	134	128	40
	San Francisco	108	92	109	353	150	103	94	143	101	87	76	103	165	125	119	51
Colorado	Denver	122	98	123	338	162	119	100	156	109	100	97	118	178	137	135	71
Connecticut	Bridgeport	128	105	130	353	170	123	106	156	115	105	100	123	186	144	142	71
	Hartford	125	101	127	338	165	122	102	159	112	103	100	121	181	140	139	74
D. C.	Washington	127	107	129	353	169	120	109	164	115	104	96	121	185	144	142	63
Florida	Jacksonville	143	130	149	406	193	128	134	189	132	117	97	134	209	167	164	47
	Miami	152	142	161	406	203	133	147	201	140	125	103	141	219	179	178	41

Note: Figures include design energy requirements for heating cooling domestic hot water fans exhaust fans heating and cooling auxiliaries elevators escalators and lighting

Note: Space is reserved in this table for restaurants and industrial buildings

itself covers not only the building envelope (DEMAND model) but also the building services systems (SUPPLY model) to give a comprehensive assessment of the actual energy consumed in a building over a period of climatic data up to a maximum of one year. It is possible to use the program to compare not only building characteristics but also system and plant options, as well as perhaps the use of different fuels.

THE PROJECT APPROACH

The use of the computer is now becoming commonplace in assessing the complex interactions between a building, its surrounding climate, its occupancy patterns and its building services system behaviour. Software has been written over the last fifteen years and reported at a series of international symposia (Ref. 4). The programs that have been developed in the US, for example, are shown in Fig. 2 (Ref. 5). Leading US energy programs are now considered to be BLAST, DOE-2 and ESP-11, all of which require large computers and complex inputs to be run effectively. In the United States attention was therefore turned to the production of a simplified procedure and the Technical Committee 4.7 of ASHRAE have supported this work (Ref. 6). Such a procedure has indeed been derived but it has shortcomings in that it does not model particularly well the transient behaviour of the building and the solar effects upon it. Thus at the commencement of our project we were faced with decisions on whether to use overseas programs or to create one suitable for use in Australia. We recognised that there was a need for a 'supply' rather than a 'demand' model so that different building services systems could be compared and different fuels could be contrasted. None of the overseas programs fulfilled the perceived Australian need for a design tool to be available for use early in the design procedure and easy to access, with well supported documentation and personal consultation services. None was in the appropriate units, and none had an Australian climatic data bank. The project undertook a brief assessment of the merits of the major US programs and concluded that DOE-2 was probably the most accessible, it being a public sector program and the best supported. The project team therefore decided to obtain DOE-2 as a back-up program and retain it for possible use in the latter stages of a building project or for comparison with the locally produced program. In summary then, the approach in this project for non-residential buildings has been:

(a) To obtain DOE-2, as a reference program.

(b) To write the BUNYIP program for a range of computers, for design office use. To rely on experience with program TEMPER in this task, adding a menu of major system and plant options.

(c) To validate BUNYIP against several buildings, including one being monitored under the IEA project in Glasgow, and several less detailed Australian buildings.

(d) To develop Bureau of Meteorology data into a suitable form for running with BUNYIP and to make it available as a climatic data bank for any other program.

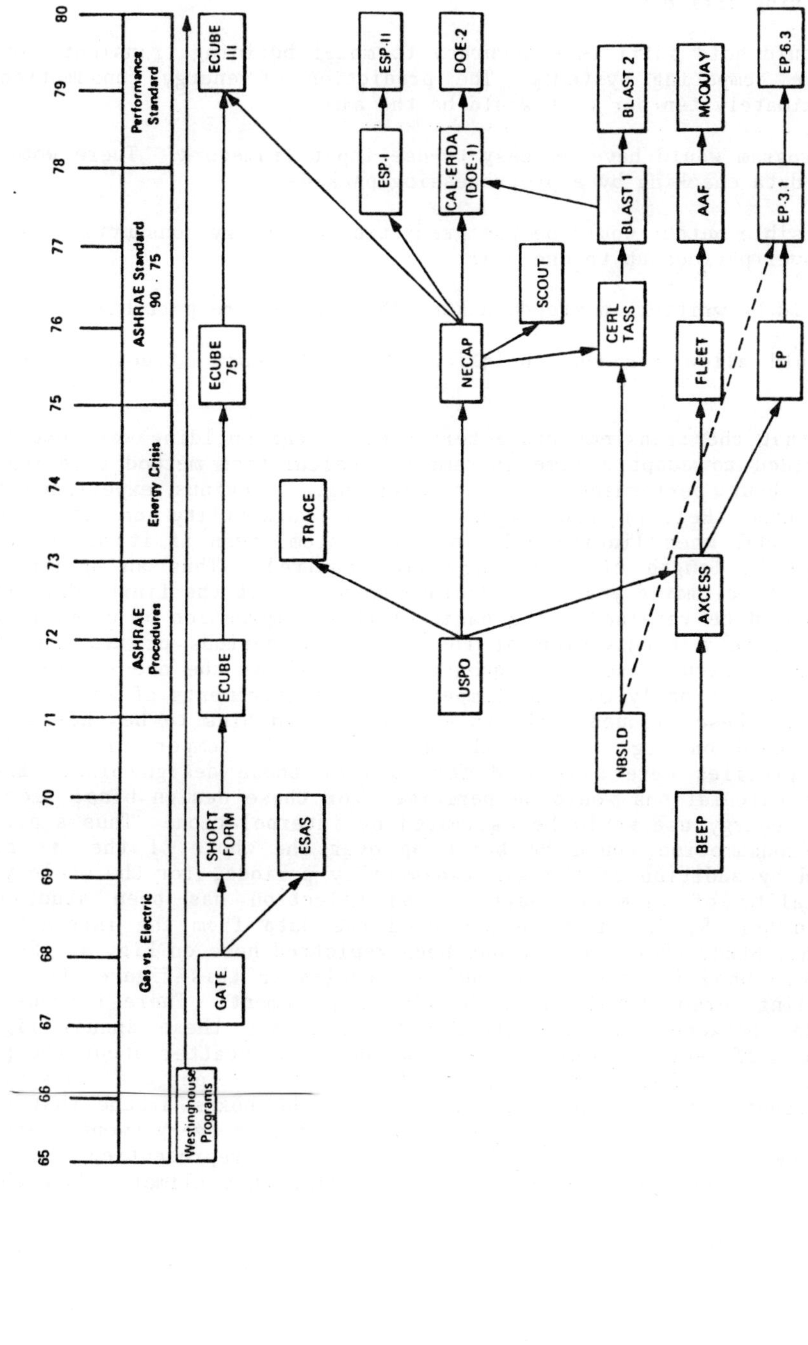

Source: Adapted from Information Provided by Francis Ferreira, Ferreira Associates, Inc.

Fig. 2 History of US energy analysis computer programs (source: Ref. 5)

PROGRAM BUNYIP

Several criteria were identified when framing the guidelines covering development of the BUNYIP package for non-residential building energy estimation. They include:

- That it should fit onto both a main frame computer or onto a small computer of the size and type which might be found in a medium sized consulting office.
- It should hold sufficient accuracy to model building transients and to compare competing systems. The prediction of energy consumption to approximately ten per cent would be the aim.
- The program would have an easy-to-use input framework. There would be rapid data checking by a preprocessing package.
- The visible output would be the prediction of energy consumption of each fuel over periods up to one year.
- It would be written in standard FORTRAN IV to ensure portability.
- It would allow for the inclusion of tariff cost structure routines later.

To ensure that the transient characteristics of the building were modelled, it was decided to adopt a time incremented calculation method covering the twenty-four hour performance of the building. Previous experience with program TEMPER (Ref. 7) had emphasised the versatility on the finite difference model when linking building and system, even if it may suffer a disadvantage in length of computing time required. Thus an option of a reduced length 'climatic year' was decided upon so that the finite difference procedure could be retained. The basis of this compression of climatic data would firstly be a subdivision of the year into periods of two months or 'seasons'. For each season a frequency chart such as the one in Fig. 3 was compiled, based on daily maximum temperature, in increments of 2°C, or bins. Inspection of these frequency charts enabled design bins to be chosen (such as those shaded on Fig. 3), and the average hourly temperature and solar irradiance profiles were developed for each of these design bins. Energy consumption calculations would be performed for these design bins; for the other bins, energy use would be estimated by interpolation. Thus a picture of energy consumption could be built up over the whole of the two month season, and by addition of the six two monthly periods, for the whole year. The criticality of climatic data period selection has been studied by Guettler in Ref. 8, in which he has used the data from the International Energy Agency Study (Ref. 9): it has been replotted here on Fig. 4. Several programs were used in this study and the results on this figure show total annual cooling versus total annual heating requirement. There is considerable scatter between all the attempts to calculate these annual figures amongst the different programs. There is much less scatter about the point marking Guettler's FRG program using different climatic periods. His single point represents an 8760 hour calculation and the boxes around that point show calculations with one day in every ten, one day in every twenty, and one day in every thirty days, i.e. one day per month, representing the whole year. It can be seen on that particular building that climatic data choice

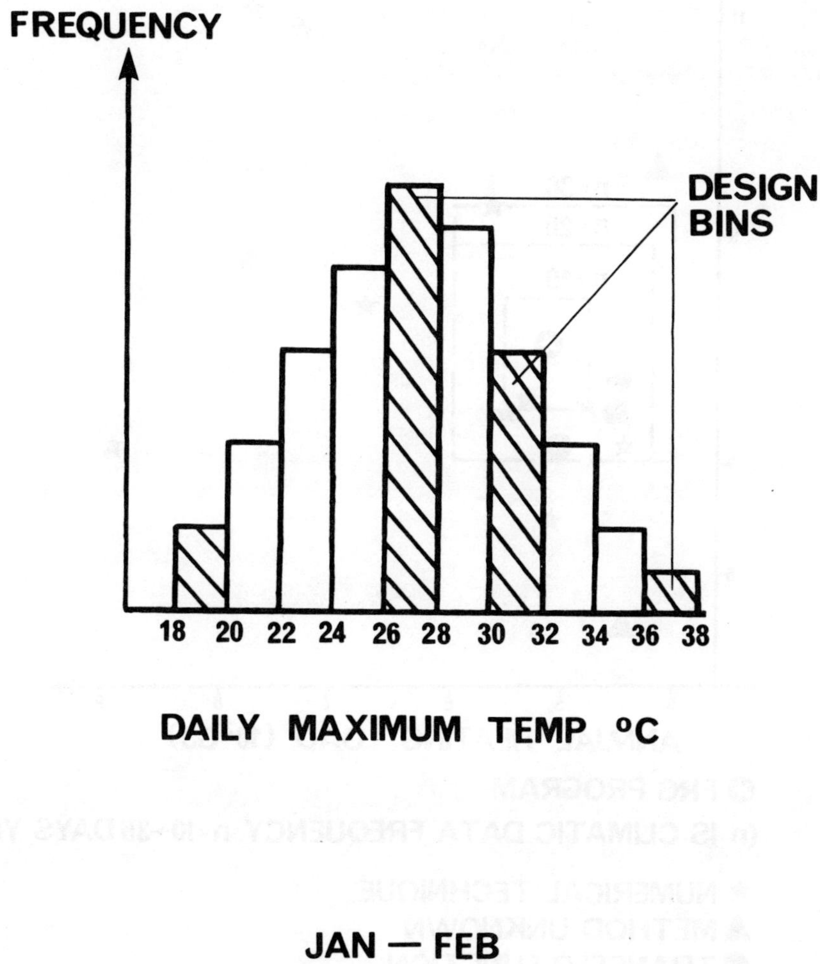

Fig. 3 Typical presentation of frequency of occurrence data for binning on the basis of daily maximum temperature

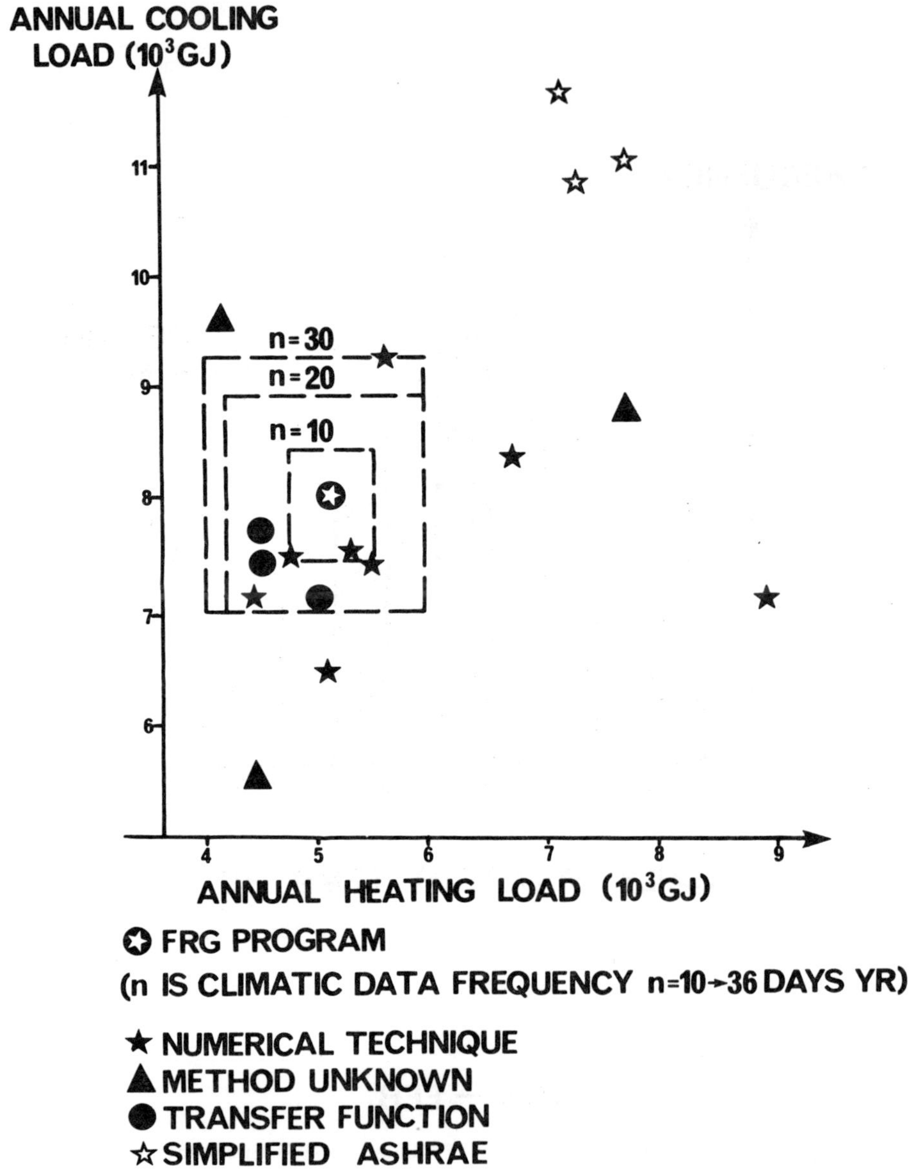

Fig. 4 Annual energy demand comparison as computed by several programs in IEA Study (Ref. 9), with climatic sensitivity deduced in Ref. 8

was less significant than was the choice of program itself to do the calculation.

The BUNYIP package consists essentially of the steps shown in Figs 5 and 6. The input phase, Fig. 5, illustrates how the climate, the building materials and the building operating schedule are input together with the systems and plant architecture. Once the data has passed the test of its acceptability it is set up in a data file to do the computation shown in Fig. 6. In this figure it can be seen that the first of the design bins of hourly data, which has been preselected, are read in for the location in question in the first box. From the input of usage levels and schedule data the program can establish time dependent parameters, such as occupancy and lighting levels, and it steps through a time loop for a maximum of three days. These days are all identical, being two repeats of the design bin day. If the user is analysing a ten hour per day, five day per week occupancy building, the first of these days corresponds to the 'Sunday', the second to a 'Monday' and the third to a 'Tuesday-through-Friday'. On the Sunday, no plant or system is operated and only the building temperature is calculated, no energy being consumed. The energy consumed on the Monday and the Tuesday is accumulated and weighted accordingly in the totalling process. If the user is analysing a twenty-four hour, seven day per week building, such as a hospital, the program only executes two days of calculation, the system and plant being on for both days. Energy consumption data for only the second day is used in the accumulation process.

The simulation of the building, zone by zone, system by system, is performed in this way for each of the design bins in a season, and for each season. The accumulated energy consumption data is stored until reported at the end of the program. No optimisation is attempted in this program.

Turning now to the way in which the program is being put into the market place; there will be an eighteen month trial and user evaluation period following completion of the project in late 1982. This will be conducted by an energy consulting firm in Melbourne and will result in training courses being run in Australian cities together with support documentation being produced. User input formatted pads will be available for this phase along with a comprehensive user manual. Upon completion of this activity the program will be released generally through the library of the Association for Computer Aided Design (ACADS) where it will be supported through that library. The Division of Energy Technology, CSIRO, will also be evaluating the program during 1983 with comparisons against data produced in an International Energy Agency project centered in Glasgow, where a four storey commercial publishing house building is being extensively monitored. Data from a small number of Australian buildings will also be available and be used to test the program.

In discussing commercialisation of the program it is worth reconsidering who should write, assemble, support and market this type of software. It would seem that the task of writing the fundamental algorithms could well fall to research organisations and universities, but clearly the assembly of these algorithms and routines into a program would more nearly be the task of the private sector in which a particular program produced by, say, a consultant, could be seen to have commercial advantage. Researchers rarely know the real requirements of the market place and it was for this reason that the present project took advantage of an ad hoc committee formed to advise the project

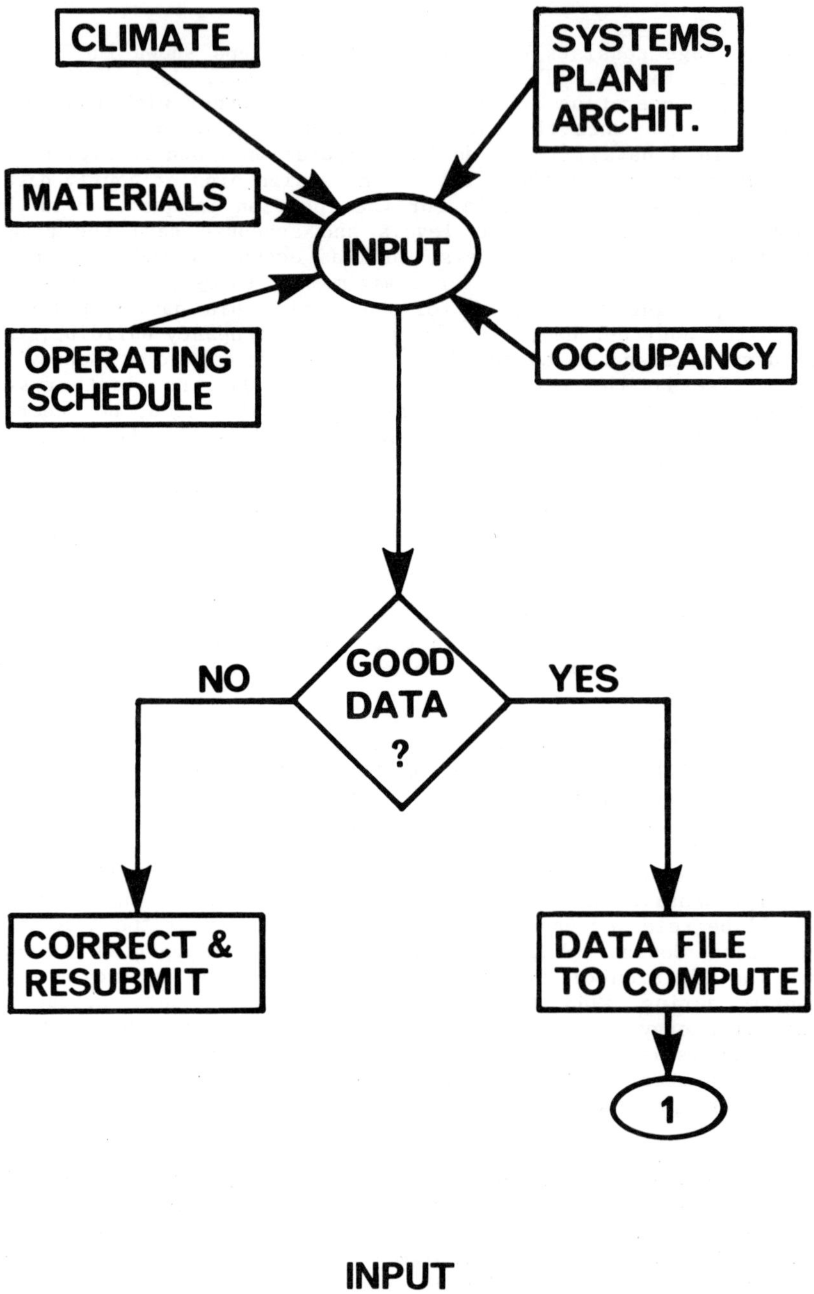

Fig. 5 Program BUNYIP: simplified input process

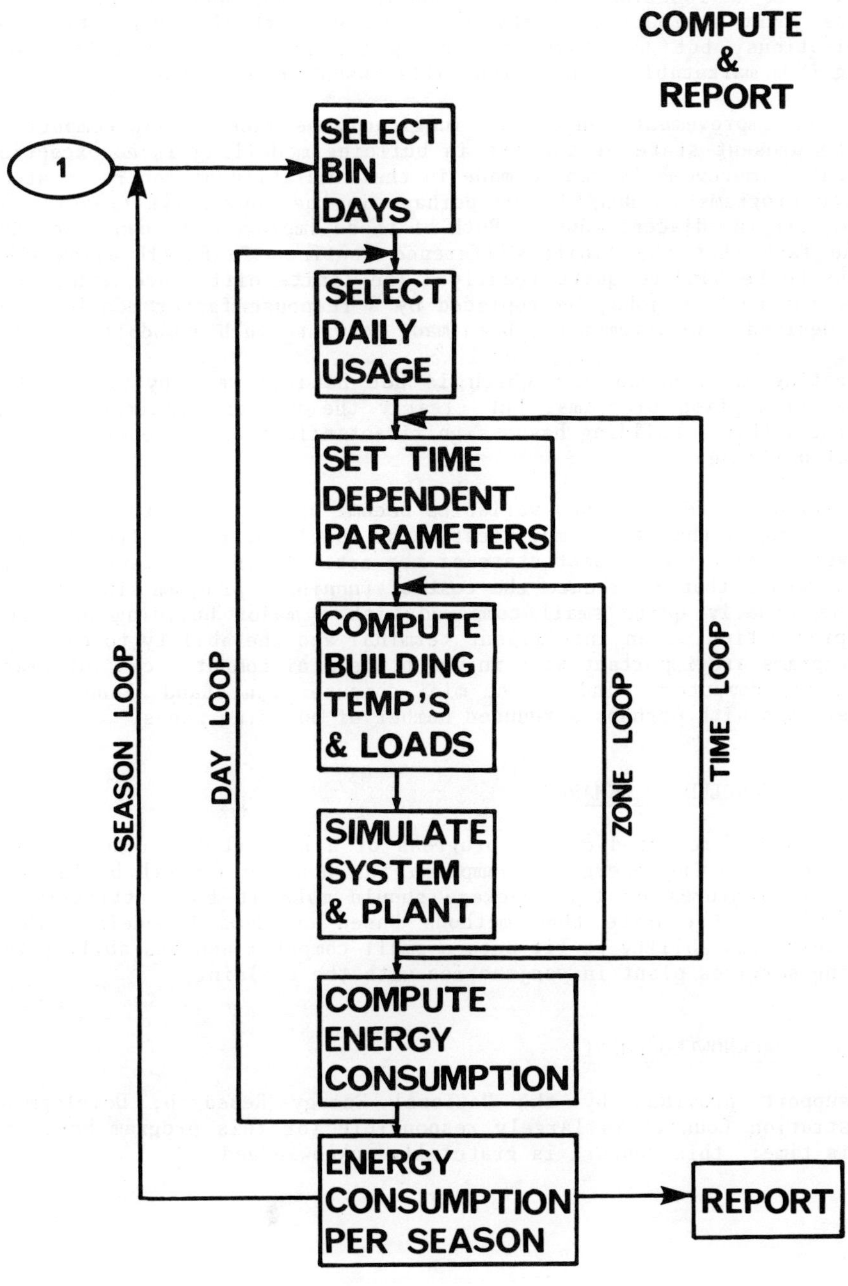

Fig. 6 Program BUNYIP: simplified compute process

team of user requirements. Clearly the support and marketing of such programs is a function for the computing bureaux and the private sector. Perhaps joint evaluations could be conducted with the originating research organisations, but the task of putting the program through its paces and making it a marketable proposition falls essentially in the private sector.

There are improvements which can inevitably be made to any computing model and the present state of the art in building modelling is no exception. In particular improvements can be made in the simulation of control systems used in such programs as BUNYIP, and perhaps in the convective transfer of heat between air in adjacent zones. Both of these improvements can take advantage of the fact that the finite difference routines in BUNYIP allow these two effects to be handled quite readily. The finite difference model in BUNYIP could, for certain jobs, be replaced by a response factor module, should it be so desired. No attempt has been made to write such a module.

Daylighting is a phenomenon which is not handled well by present building energy consumption programs, but clearly the use and optimisation of daylighting within a building has a significant effect on the energy consumption in that building.

As individual effects of key variables become better understood and documented it is hoped that the simplification of models will be a fruitful area for improvement over the present state of the art. It is important to facilitate access rather than to reduce the cost of running a program since the running cost is usually quite small compared with a major building project cost. Data preparation on an intelligent terminal and the ability to execute editing programs are important ways in which the mini computer can interface with main frame computer. Indeed the mini computer can stand alone for certain smaller jobs with perhaps a reduced number of building zones.

CONCLUDING REMARKS

This paper has reported on the progress of a project to produce a means of estimating building energy consumption for non-residential buildings. Two particular features of this package should make it both attractive to the user and more accurate than methods based on demand models. They are, separately, its ability to fit onto a small computer and its ability to model building services plant in conjunction with the building.

ACKNOWLEDGEMENT

The support provided by the National Energy Research, Development and Demonstration Council is largely responsible for this program being created at this time; this support is gratefully acknowledged.

REFERENCES

(1) Energy Conservation in Buildings. National Energy Advisory Committee, Report No. 13 (1980).

(2) Personal communications with several consulting engineers.

(3) US Government: Federal Register, November 28, 1979. Energy Performance Standards for New Buildings : Proposed Rules.

(4) Proceedings of the International Conference on the Use of Computers for Environmental Engineering Related to Buildings Washington, USA (1970), Paris, France (1974) and Banff, Canada (1978).

(5) Arthur D. Little Inc. Building Energy Analysis Computer Programs with Solar Heating and Cooling System Capabilities. Research Project 1269, May 1979.

(6) T. Kusuda, A Simplified Energy Calculation Procedure for Use in Energy Conservation Standards Activities. ASHRAE Technical Committee 4.7 Report (1979).

(7) M.J. Wooldridge, Air-conditioning load calculations using computer program TEMPER. Aust.Refrig.Air Condit. Heat., 26(12), 26 (1973).

(8) G. Guettler, Test Reference Year Versus Reference Day Concept. Paper presented at IEA Conf. on New Energy Conservation Technologies, West Berlin, April 1981. Springer Verlag, 1981.

(9) J. Cockroft, IEA - Results of Real Building Comparisons - Avonbank Collins. Paper presented at IEA Conf. on New Energy Conservation Technologies, West Berlin, April 1981. Springer Verlag, 1981.

A Thermal Design Tool for the Sketch Design Stage

S.V. Szokolay
Director, Architectural Science Unit,
University of Queensland

INTRODUCTION

> "The climate...presents a challenge to the architect, not satisfied with substituting mechanical equipment for good design."

This was written by H.J. Cowan (1) in an editorial of the Architectural Science Review some twenty-three years ago (Nov. 1959). In those years this was largely a question of attitude, of philosophy, of ethics. It took the energy crisis of the '70s to bring home the point and turn it into a practical, even financial question.

Most architects would - at best - consider the climate only in qualitative terms, when designing a building. And they can't really be blamed, as the simple manual calculation methods are both crude and cumbersome and to employ computer simulation is a major undertaking. We felt that a tool was needed, which could be usefully and easily employed at the sketch design stage.

We set ourselves the following specifications for such a tool:

- a mini- (possibly a micro-) computer should be able to support it,
- it should be inexpensive to run it (in money or in time),
- it should be an interactive program, simple enough for use by people with little or no computing experience,
- an uninitiated person should be able to learn to use it in less than a day,
- the input data requirement should be minimal, with as little pre-calculation needed as possible,
- the input method should be simple and quick,
- any design parameter should be capable of being changed, quickly, whilst keeping all other data,
- the output should be concise and immediately understandable,
- precision of the results in absolute terms is less important than the correctness of comparative effects of changes.

CHOICE OF METHOD

The various approaches to building thermal response simulation were first reviewed. Analytical techniques, such as the finite difference method, which trace the heat flow from layer-to-layer, or the various solutions to resistance-capacitance network analogies (eg. the matrix method) are far more accurate than the description of the building, or indeed, than building construction itself. The response factor method facilitates the simulation of the performance of a given building through a large number of days of climatic data, but every change to the building would necessitate the re-calculation of response factors. Thus the constant (the response factor) is the one we would need to have as a variable.

The "admittance procedure" of the UK Building Research Establishment (2,3) seemed to be the most obvious choice. It is conceptually simple and has been used in manual calculations for many years. It is based on the calculation of the 24-hour mean heat flow through an element:

$$\bar{Q} = A \cdot U\,(\bar{T}_s - T_i)$$

(where $\bar{T}_s$ is the 24-hour mean sol-air temperature of the outside surface and T_i, the indoor temperature, is taken as constant), followed by the calculation of the deviation from this mean for each hour θ:

$$\tilde{Q} = A \cdot U \cdot \mu (T_{s(\theta-\phi)} - \bar{T}_s)$$

where μ = decrement factor

ϕ = time-lag

$T_{s(\theta-\phi)}$ = sol-air temperature of surface, ϕ hours previously

These are the principles. In the actual method T_e (environmental temperature) is used rather than T_i and it is the unknown.

The net conduction heat flow due to air temperature difference, over the 24-hour cycle is taken as zero, thus in the absence of other gains $\bar{T}_e = \bar{T}_o$ (where T_e is indoor environmental temperature and T_o is outdoor air temperature). The 'extra' gains are then calculated as

$$\bar{Q} = \bar{Q}_s + \bar{Q}_g + \bar{Q}_i \qquad \text{... 1.)}$$

ie. the solar excess gain through solid elements + solar gains through glass + internal gains (see Appendix 1). This 'extra' gain must be dissipated by conduction and ventilation, driven by the $\bar{T}_e - \bar{T}_o$ differential. From this the indoor mean environmental temperature can be found:

$$\bar{T}_e = \bar{T}_o + \frac{\bar{Q}}{\Sigma(A \cdot U) + C_v} \qquad \text{... 2.)}$$

where C_v = "ventilation conductance", for small ventilation rates taken as $C_v = c \cdot \rho \cdot v = 1200\ v = 0.33\ N \cdot V$ (v = ventilation rate, V = volume of room, N = number of air changes per hour)

but for larger ventilation rates it should be compensated by the "air - environmental" conductance, taken as 4.5 W/m^2K, thus

$$C = \frac{4.5\ \Sigma A \cdot c \cdot \rho \cdot v}{4.5\ \Sigma A + c \cdot \rho \cdot v} \quad (c = \text{specific heat capacity}, \ \rho = \text{density of air}) \qquad \ldots\ 3.)$$

In the next step the deviation from the mean heat gain rate at time θ is calculated ($\tilde{Q}_\theta$), as a sum of all component heat flows (see Appendix 1.). This deviation in heat gain is either dissipated through ventilation or stored in the fabric. The rate of heat entry into constructional elements is the <u>admittance</u> (Y) and for the whole space, the area-weighted total <u>admittance</u>: $\Sigma(A \cdot Y)$. From this the deviation in internal environmental temperature at time θ can be found as

$$\tilde{T}_{e(\theta)} = \frac{\tilde{Q}_\theta}{\Sigma(A \cdot Y) + C_v} \qquad \ldots\ 4.)$$

A somewhat more involved calculation method, which allows for variable ventilation rates, has been proposed by Petherbridge (4) and adopted for the purposes of our program (see Appendix 2.).

SYSTEM STRUCTURE

The method is derived from Danter's (5) work on harmonic analysis (which in turn was based on the work of Mackey and Wright, (6,7)) hence the program was named HARMON. Perhaps more important than HARMON itself is the structure of the whole package (8). The package consists of HARMON itself, plus three supporting programs: BUILD, CONMAT and LOCAT (Fig. 1.).

BUILD is used to create a building data file, accepting inputs (from the keyboard) of dimensions, orientation, tilt and constructional designation of each building surface, shading devices, if any, as well as ventilation rates and internal gains, which can be either constants or hourly values. The location of the building must also be specified.

When HARMON is run, it first copies over the BUILD.DAT file. It then reads from CONMAT.DAT the properties of all constructional elements designated. CONMAT.DAT contains data for all required fabric properties (U-value, time-lag, decrement factor, admittance, etc.) for a large number of constructional elements. If the designated construction is not found, the program CONMAT is called up and used to create new entries. For such a new element the thickness and material designations must be input for each layer. Then the program will read from the MATERI.DAT file the basic material properties (density, specific heat

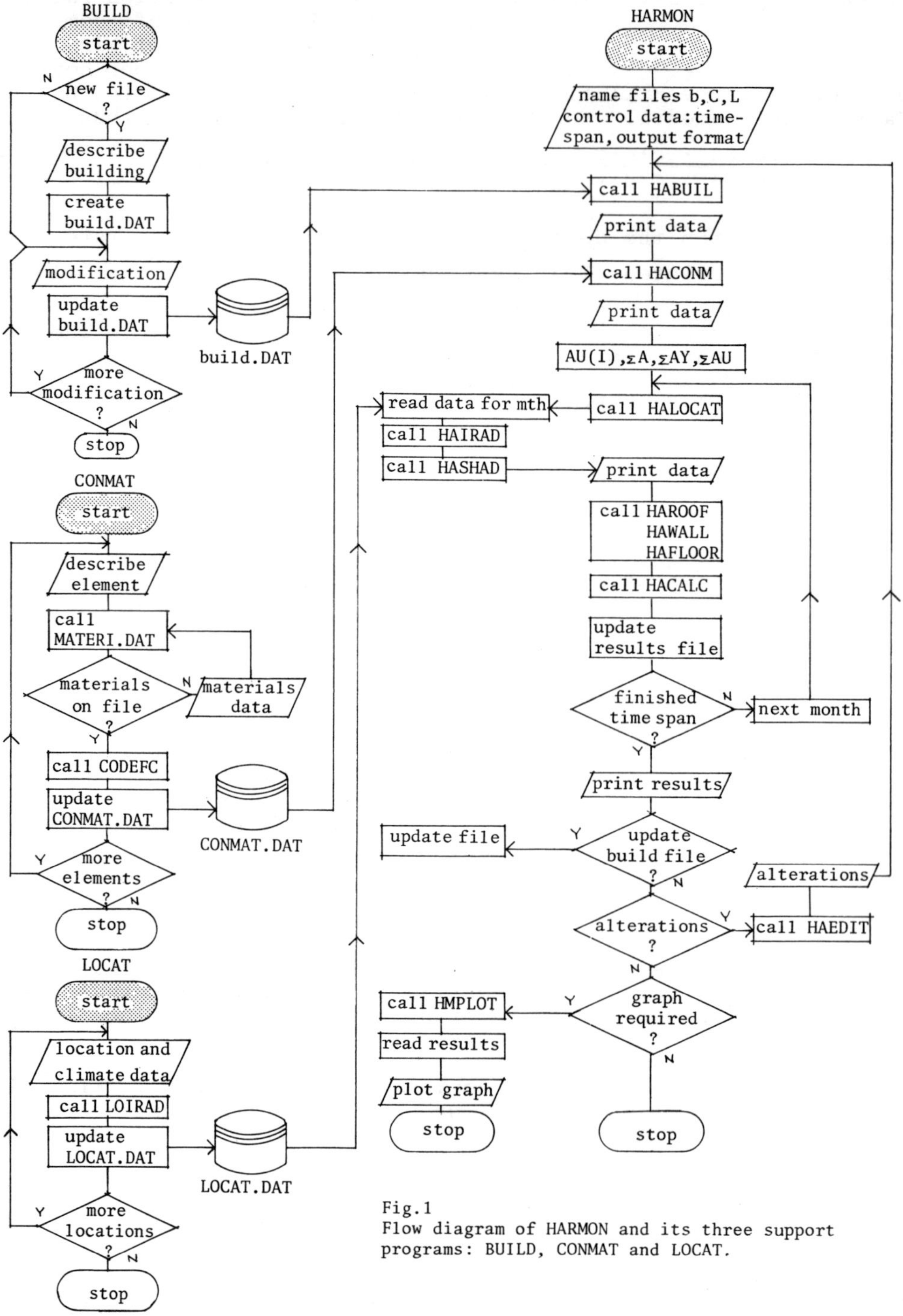

Fig.1
Flow diagram of HARMON and its three support programs: BUILD, CONMAT and LOCAT.

capacity, conductivity) and subroutine DECFAC will calculate the decrement factor and all the required fabric properties. (As it became part of the CONMAT program, it was re-named CODEFC.) Facility exists also for inputting new materials data.

This subroutine uses a complex matrix inversion algorithm and is indeed the crucial part of the package from the point of view of heat transfer theory (see Appendix 3.).

HARMON then calls up the LOCAT.DAT file and reads the latitude and climatic data for the location. If this location is not yet on file, the program LOCAT will be called up to create a new entry. This will normally require the input of hourly values of 14th, 50th and 86th percentile temperatures for one day of each month (see Appendix 4.), as well as hourly average irradiance data.

Where only the daily total irradiation value is available, subroutine LOIRAD will estimate the diffuse component and the hourly values of both global and diffuse irradiance (see Appendix 5.).

Within the program HARMON, a subroutine called HAIRAD will calculate hourly irradiance values for each building surface. If shading devices are used, subroutine HASHAD will calculate their effect at hourly intervals. The main part of the calculations is executed by subroutines HAROOF, HAWALL, HAFLOR and HACALC.

The HARMON run is started with stating the required time-span of the analysis. In the simplest case this could be one winter day and one summer day, but it can be one typical (50th percentile) day of each month, or even three days (14th, 50th and 86th percentile) for each month.

For output, a whole range of options is available. In the simplest case, (when one winter and one summer day are examined) the output may be only the maximum and minimum indoor temperature for each day. The next option is to print out hourly values of indoor temperatures (possibly both environmental and air temperatures). If a full year analysis is carried out, the printout can be obtained in the form of a 12 months x 24 hours matrix. Any results can be stored in a designated file and the program HMPLOT can be called up to obtain a graphic output of this file.

MAN-MACHINE INTERACTION

Our aim was to create an interactive program. So, when the machine gives the result for one design hypothesis, the designer must judge whether it is satisfactory or not. If yes, then the run is finished. If not then, as the computer asks: "any alterations?", the designer can answer "yes". In this case HARMON returns to the copied-over BUILD.DAT file and will ask what is to be changed. The designer can change one, or a whole series of variables at a time. In all other respects the building remains as earlier defined. Then the program is re-run and the results can be compared with the earlier ones. The program can be re-run any number of times, quite quickly, as the chores of the input are obviated. The new input for each run is only the changed variable.

Any such changes are executed on the copy of the BUILD.DAT file within HARMON. The original BUILD.DAT file is preserved within the BUILD program. After each run the program will ask: "is BUILD file to be updated?". If the answer is yes, then the latest version of the building file will be copied over from HARMON into BUILD, to replace the original. The copy kept in HARMON can however be further altered and re-run.

This gives a great flexibility to the designer. Say, we did six runs and the results were consistently improving. The last run looks the best, but we want to try a set of quite different changes. We can copy back and preserve the last changed version and examine the effects of further changes by any number of new runs.

In this "man-machine interactive system" the weakest point is the interpretation of the machine's output by the designer. He or she is to decide whether a solution tested is good or bad, whether it is better or worse than previous solutions. He or she is then to decide what further modifications should be examined, which are likely to lead to improvements.

To put it into the jargon of information science (9):

- the designer is striving for a "goal set", G
- the designer is generating solutions (set S) which are expected to be part of set G
- if U is the universal set of all possible solutions, then both sets S and G are sub-sets of U
- the machine evaluates these solutions, to check whether they are in fact parts of set G.

There are several problems in this. Firstly, it must be recognised that this is not optimisation, but a heuristic process. It consists of a series of cycles of solution generation and evaluation. It may produce a 'good' solution, but it is unknown whether it is the 'best'. Secondly, there are no precise evaluation criteria, thus the computer cannot complete the evaluation, it can only contribute to the evaluation process, which must then be completed by human judgement. Thirdly, even if there were a precise definition of the 'goal set' from the thermal point-of-view, the building design as a whole would remain an ill-defined problem. The designer must apply a large number of criteria, many of which are qualitative only.

EVALUATION METHODS

Coming back to the definition of the 'goal set' from the thermal point-of-view only: we can

(a) specify the maximum permissible (or acceptable) energy consumption for heating and cooling, for keeping the building within a stated narrow range of temperatures, or

(b) we can compare the predicted indoor temperature profile for the non-conditioned building with

 (i) the outdoor temperature profile

 (ii) with the 'comfort zone' band of temperatures

 (iii) with the temperature profile of a similar conventional building, or indeed with all three (Fig. 2).

We decided against using approach (a), as the specification of permissible energy consumption itself involves certain arbitrary judgements and - to our knowledge - it has only been attempted so far in the Californian building energy code (10). Choosing approach (b) we allowed for several options.

(1) Plots of outdoor and predicted indoor hourly temperatures are produced for one winter and one summer day. On the same graph we plot the band of 'comfort range' defined by the function proposed by Auliciems (11):

the neutrality temperature is $T_n = 17.6 + 0.31\ \overline{T}_o$

where $\overline{T}_o$ is the mean temperature of the month.

The limits are then taken as

upper: $T_u = T_n + 2\ ^{0}C$ lower: $T_\ell = T_n - 2\ ^{0}C$

This band is indicative only. We set no precise target, but plotting the results of successive runs on the same graph, we can select the one nearest to the comfort zone. When we see what a conventional building can produce and when we gradually apply 'reasonable' improvements, we see what is possible, so our goal-set definition in fact emerges only as we continue the search cycles.

The graphic output here is essential, as it facilitates quick, visual comparisons.

(2) The purpose of even quicker, but still meaningful machine-to-man communication is served by the use of the THI (thermal habitability index), developed by Woolard (12). This is a single-figure index, calculated by subroutine THCAL, by the following method:

four sums are found

- for winter, from response to the 14th percentile climatic data:

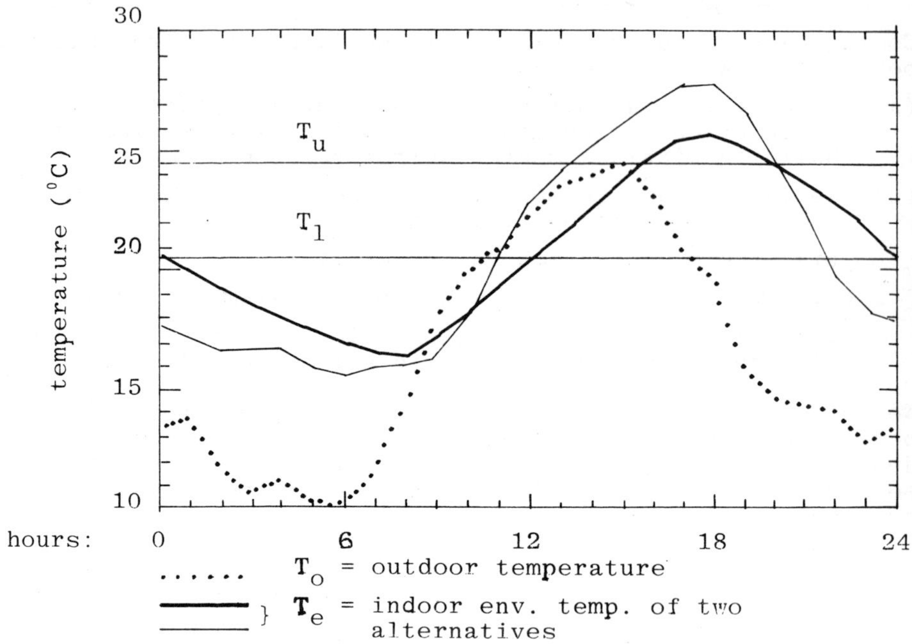

Fig. 2 HARMON output: temperature profiles

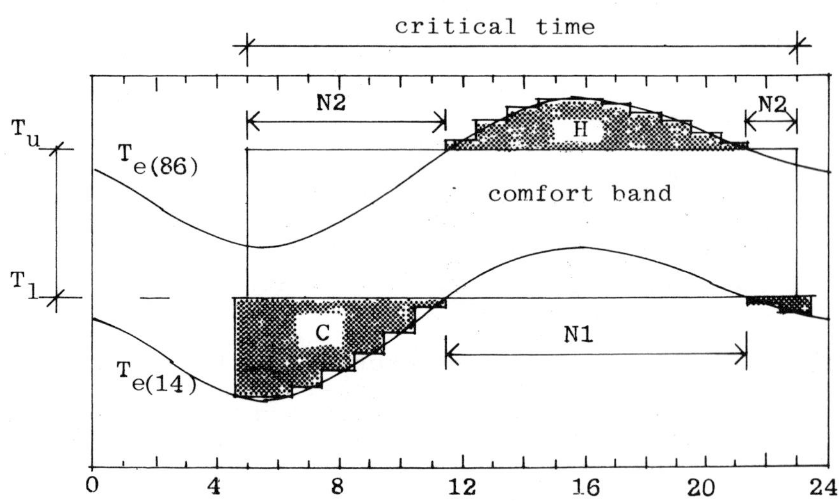

Fig. 3 Quantities used for Thermal Habitability Index

N1 = number of hours when $T_e > T_\ell$

C = Kelvin-hours when $T_e < T_\ell$

- for summer, from response to the 86th percentile climatic data:
  N2 = number of hours when $T_e < T_u$

In both cases the time-base can be the full 24 hours (for continuously occupied buildings) or a specified time span, corresponding to the period of occupation (See Fig. 3 for interpretation).

The indices are then produced:

$$THI_{cold} = \frac{N1}{N1 + C} \cdot 100$$

$$THI_{hot} = \frac{N2}{N2 + H} \cdot 100$$

$$THI_{tot} = \frac{N1 + N2}{N1 + N2 + C + H} \cdot 100$$

The value of 100 would indicate that there is no thermal problem and the smaller the value, the greater the thermal deficiency.

THI indices can be produced from a run for one winter and one summer day, or from a full twelve months (ie twelve representative days) run. Each day is run twice, once with the 14th and once with the 86th percentile climatic data. For the quickest machine-to-man communication all other outputs may be suppressed, only the three THI values printed out. The solution giving THI values nearest to 100 is the best.

(3) The THI is not only concise, but it does include all that is important in characterising the thermal performance of a design hypothesis. So, if thermal performance were the only concern of the designer, there would be no need for any other output. However, in many instances (eg. in ref. 13) there may be a need to obtain more information such as when the under- or overheating occurs, so that it could be compared with occupation patterns, which may even have planning implications.

A useful form of obtaining such information from the machine is a 12 months x 24 hour matrix. (In this case the full twelve months' run will be necessary.) We would produce two such matrices, one for the 14th and one for the 86th percentile climatic data. The lower comfort limit contour would then be drawn on the former, the upper one on the latter, by interpolation between the grid-point values (Fig. 4). If such plots are obtained on tracing paper or film, then the results of successive runs will be readily comparable.

86th %-ile, max. vent. with overheated period

	JULY	AUG.	SEP.	OCT.	NOV.	DEC.	JAN.	FEB.	MAR.	APR.	MAY	JUNE
0	16.7	17.7	20.2	23.0	25.1	26.2	27.0	26.6	25.2	23.0	20.1	17.5
1	16.6	17.6	20.1	22.9	25.0	26.1	26.9	26.5	25.1	22.9	20.0	17.4
2	16.3	17.3	19.8	22.6	24.7	25.8	26.6	26.2	24.8	22.6	19.7	17.1
3	15.9	16.9	19.4	22.2	24.3	25.4	26.3	25.9	24.5	22.2	19.4	16.8
4	15.5	16.4	18.9	21.8	23.9	25.0	25.9	25.5	24.1	21.8	19.0	16.4
5	15.2	16.0	18.5	21.4	23.5	24.7	25.6	25.2	23.7	21.4	18.7	16.0
6	15.0	15.8	18.3	22.8	25.7	27.0	27.8	26.9	24.7	21.3	18.5	15.9
7	17.1	18.0	20.8	24.0	26.8	28.2	29.0	28.0	26.1	23.5	20.4	17.6
8	19.7	20.3	23.0	26.0	28.6	30.1	30.8	29.7	28.2	25.9	22.4	19.9
9	21.8	22.6	25.3	28.3	30.5	31.8	32.4	31.8	30.4	28.1	24.5	22.0
10	23.7	24.7	27.4	30.3	32.4	33.4	34.1	33.6	32.2	30.1	26.4	23.8
11	25.4	26.7	29.4	32.2	34.2	35.1	35.8	35.2	33.9	32.0	28.1	25.6
12	26.5	28.0	30.7	33.5	35.4	36.3	37.0	36.4	35.1	33.3	29.3	26.7
13	27.5	29.0	31.8	34.4	36.4	37.2	37.8	37.2	36.0	34.2	30.1	27.6
14	27.5	29.1	31.9	34.6	36.5	37.4	37.9	37.3	36.0	34.2	30.2	27.6
15	26.7	28.3	31.1	33.8	35.8	37.0	37.4	36.6	35.3	33.5	29.4	26.9
16	25.3	26.8	29.6	32.4	34.7	36.0	36.4	35.2	33.8	32.0	28.0	25.5
17	23.3	25.1	27.7	30.6	33.3	34.5	35.0	33.7	32.0	30.2	26.3	23.6
18	22.1	23.5	26.0	29.3	32.1	33.3	33.8	32.5	30.7	28.6	25.1	22.6
19	21.0	22.3	24.8	27.5	29.5	30.6	31.2	30.6	29.3	27.4	24.1	21.5
20	19.5	20.8	23.3	26.1	28.1	29.2	29.9	29.3	27.9	25.9	22.7	20.1
21	18.4	19.6	22.1	25.0	27.1	28.2	28.9	28.3	26.9	24.8	21.7	19.1
22	17.7	18.8	21.3	24.1	26.2	27.3	28.1	27.6	26.2	24.0	21.1	18.5
23	17.0	18.0	20.5	23.3	25.4	26.5	27.3	26.8	25.4	23.3	20.4	17.8

14th %-ile, 0.5 ach/h with underheated period

	JAN.	FEB.	MAR.	APR.	MAY	JUN.	JUL.	AUG.	SEP.	OCT.	NOV.	DEC.
0	22.5	22.7	21.1	19.6	15.2	13.2	11.5	13.4	15.9	18.9	21.1	22.7
1	22.5	22.6	21.0	19.6	15.1	13.1	11.4	13.3	15.9	18.8	21.1	22.6
2	22.3	22.4	20.8	19.3	14.8	12.8	11.1	13.0	15.6	18.6	20.8	22.4
3	22.0	22.1	20.5	19.1	14.5	12.5	10.8	12.6	15.3	18.3	20.6	22.2
4	21.8	21.9	20.2	18.8	14.2	12.2	10.4	12.2	14.9	18.0	20.3	21.9
5	21.5	21.6	20.0	18.5	13.9	11.9	10.0	11.9	14.6	17.7	20.0	21.7
6	22.4	22.8	20.5	18.4	13.7	11.7	9.9	11.7	14.4	19.0	21.4	23.1
7	23.5	23.4	21.2	20.0	15.3	13.1	11.4	13.4	16.4	19.7	22.7	24.5
8	25.2	25.0	22.9	21.8	16.7	14.6	12.9	15.2	18.6	21.8	24.7	26.5
9	26.7	26.8	24.6	23.7	18.7	16.5	15.0	17.3	20.7	23.8	26.4	28.2
10	27.8	28.3	25.8	25.3	20.4	18.2	16.8	19.1	22.5	25.5	28.0	29.4
11	29.2	29.4	27.3	26.9	22.2	19.9	18.7	21.0	24.4	27.1	29.5	30.9
12	30.2	30.4	28.4	28.0	23.5	21.1	20.1	22.4	25.7	28.3	30.6	31.9
13	30.8	31.1	29.1	28.8	24.3	21.9	21.1	23.3	26.6	29.1	31.3	32.6
14	30.9	31.2	29.2	28.8	24.4	22.0	21.2	23.5	26.7	29.2	31.4	32.7
15	30.5	30.7	28.7	28.1	23.8	21.4	20.6	22.8	26.0	28.6	30.8	32.3
16	29.6	29.6	27.6	26.8	22.5	20.1	19.3	21.4	24.6	27.2	29.6	31.2
17	28.3	28.2	26.2	25.2	21.0	18.6	17.7	19.8	22.7	25.5	28.1	29.6
18	27.6	27.6	25.6	24.3	20.3	18.2	17.1	18.9	21.5	24.6	27.1	28.6
19	26.0	26.2	24.6	23.5	19.4	17.3	16.0	17.9	20.5	23.1	25.2	26.6
20	24.9	25.0	23.4	22.1	17.9	15.8	14.5	16.4	19.0	21.7	23.8	25.3
21	24.1	24.2	22.6	21.3	17.1	15.0	13.6	15.4	17.9	20.8	23.0	24.5
22	23.5	23.7	22.1	20.7	16.4	14.4	12.8	14.7	17.2	20.1	22.3	23.8
23	22.8	22.4	21.3	19.9	15.5	13.5	11.9	13.7	16.3	19.1	21.3	22.9

Fig. 4 HARMON output matrices of indoor T_e (^{0}C)

CONCLUSIONS

Computers are penetrating architectural practice at an increasing rate, mainly at two levels:

- office administration and accounting
- drawing and specification production.

The most important role of computers - in our view - is to aid and improve the quality of design. This level is the least developed, probably because it is the most difficult. A recent review by the Architects' Journal used the label 'Technics' for this third area of computer applications, in a rather disparaging sense. Justifiably so, as most small packages available are computerised versions of manual calculation methods, such as steady-state heat loss calculations, or the lumen method of general lighting design. Large, all-embracing computer-based design systems are in their infancy and to many, they are off-putting and even frightening.

The package described here is not more than - perhaps - a small contribution to a rather important aspect of architectural design. It is a tool, only slightly above the level of what has been described as 'technics'. It will not encroach on the 'designer's freedom' (as some of the large packages under development are feared to do), but - if used intelligently, in a way indicated by Fig. 5, it should lead to better designs, at least from the thermal/climatic point-of-view.

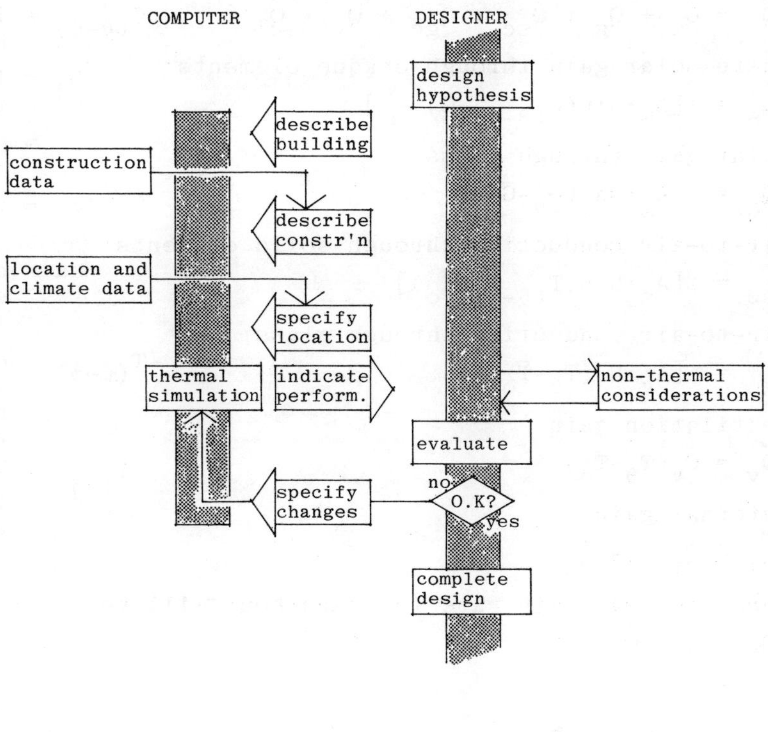

Fig. 5

APPENDIX 1

ALGORITHMS FOR THE CALCULATION OF COMPONENT HEAT FLOWS

A. 'Extra' Daily Mean Heat Flows

Mean solar gain must be calculated for each surface separately, then added. First, for opaque (solid) surfaces:

$$\bar{Q}_s = \Sigma(A_s \cdot U \frac{\alpha\bar{G}-E}{f})$$

E = long wave emission (W/m^2), if unknown, use 90 W/m^2 for roofs, zero for walls.

then for glass (transparent) surfaces:

$$\bar{Q}_g = \Sigma(\bar{G} \cdot A_g \cdot s)$$

where A = area of each element (m^2), A_s solid, A_g glass

U = transmittance of each element (W/m^2K)

$\bar{G}$ = 24-hour mean horizontal irradiance (W/m^2) on each surface

α = absorptance of each surface

f = surface conductance of each surface

s = solar gain factor

The mean internal gain rate is found by adding the products of all component gain rates and their time of use and dividing the sum by twenty-four:

$$\bar{Q}_i = \frac{\Sigma(Q_i \cdot h)}{24}$$

B. Deviation of Component Heat Flows from the Mean at Time θ

$$\tilde{Q}_\theta = \tilde{Q}_s + \tilde{Q}_g + \tilde{Q}_{sc} + \tilde{Q}_{gc} + \tilde{Q}_v + \tilde{Q}_i$$

where solar gain through opaque elements:

$$\tilde{Q}_s = \Sigma[A_s \cdot U \cdot \mu(G_{(\theta-\phi)}-\bar{G}) \frac{\alpha}{f}]$$

solar gain through glass:

$$\tilde{Q}_g = \Sigma[A_g \cdot sa\ (G_\theta-\bar{G})]$$

air-to-air conduction through solid elements:

$$\tilde{Q}_{sc} = \Sigma[A_s \cdot U \cdot \mu(T_{(\theta-\phi)}-\bar{T}_o)]$$

air-to-air conduction through glass:

$$\tilde{Q}_{gc} = \Sigma[A_g \cdot U\ (T_\theta-\bar{T}_o)]$$

ventilation gain:

$$\tilde{Q}_v = C_v(T_\theta-\bar{T}_o)$$

internal gain:

$$\tilde{Q}_i = Q_{i(\theta)} - \bar{Q}_i$$

$G_{(\theta-\phi)}$ = irradiance ϕ hours before

G_θ = irradiance at time θ

μ = decrement factor

ϕ = time lag

sa = alternating solar gain factor

$T_{(\theta-\phi)}$ = outdoor temperature hours previously

$T_{(\theta)}$ = outdoor temperature at time θ

Then the heat gain rate for each hour will be:

$$Q_\theta = \bar{Q} + \tilde{Q}_\theta$$

APPENDIX 2

ALGORITHM FOR VARIABLE VENTILATION RATE

From Eq. 2 (p.2): $\bar{Q} = [\Sigma(A \cdot U)+C_v]\ (\bar{T}_e-\bar{T}_o)$

From Eq. 4 : $\tilde{Q}_\theta = [\Sigma(A \cdot Y)+C_v]\ \tilde{T}_{e(\theta)}$

as $Q_\theta = \bar{Q}+\tilde{Q}_\theta$, we can write

$$Q_\theta = [\Sigma(A \cdot U)+C_v](\bar{T}_e-\bar{T}_o)+[\Sigma(A \cdot Y)+C_v]\ \tilde{T}_{e(\theta)}$$

dividing all terms by $\Sigma(A \cdot Y)+C_v$:

$$\frac{Q_\theta}{\Sigma(A \cdot Y)+C_v} = \frac{[\Sigma(A \cdot U)+C_v]\ (\bar{T}_e-\bar{T}_o)}{\Sigma(A \cdot Y)+C_v} + \tilde{T}_{e(\theta)}$$

Q_θ and C_v will each have twenty-four known values. Summing and averaging over twenty-four hours:

$$\frac{1}{24}\Sigma\frac{Q_\theta}{\Sigma(A \cdot Y)+C_v} = \frac{1}{24}\Sigma\frac{[\Sigma(A \cdot U)+C_v](\bar{T}_e-\bar{T}_o)}{\Sigma(A \cdot Y)+C_v} + \frac{1}{24}\Sigma\tilde{T}_{e(\theta)}$$

The last term, by definition, is zero. $(\bar{T}_e-\bar{T}_o)$ is constant.

The $\frac{1}{24}$ terms cancel out.

$$\Sigma\frac{Q_\theta}{\Sigma(A \cdot Y)+C_v} = (\bar{T}_e-\bar{T}_o)\ \Sigma\frac{\Sigma(A \cdot U)+C_v}{\Sigma(A \cdot Y)+C_v}$$

From which $\bar{T}_e$ can be expressed: if for the last equation we write

$$A = (\bar{T}_e-\bar{T}_o)\ B$$

then

$$A = \bar{T}_e\ B - \bar{T}_o\ B$$

$$\bar{T}_e = \frac{\bar{T}_o B + A}{B}.$$

Then all twenty-four values of $\tilde{T}_{e(\theta)}$ can be found from

$$\tilde{T}_{e(\theta)} = \frac{Q_\theta-[\Sigma(A \cdot U)+C_v]\ (\bar{T}_e-\bar{T}_o)}{\Sigma(A \cdot Y)+C_v}$$

and for each hour

$$T_{e(\theta)} = \bar{T}_e+\tilde{T}_{e(\theta)}$$

APPENDIX 3

ALGORITHM FOR CALCULATING THERMAL CONSTANTS OF CONSTRUCTIONAL ELEMENTS

Milbank and Harrington-Lynn (3), referring to Jakob (14) and Pipes (15) suggest the following matrix expression for linking temperature and heat flow cycles in multilayer walls:

$$\begin{bmatrix} t_o \\ q_o \end{bmatrix} = \begin{bmatrix} 1 & R_{so} \\ 0 & 1 \end{bmatrix} \times \begin{bmatrix} E & F \\ G & H \end{bmatrix} \times \begin{bmatrix} 1 & R_{si} \\ 0 & 1 \end{bmatrix} \times \begin{bmatrix} t_i \\ q_i \end{bmatrix}$$

where

$$\begin{bmatrix} E & F \\ G & H \end{bmatrix} = \begin{bmatrix} A_1 & B_1 \\ D_1 & A_1 \end{bmatrix} \times \begin{bmatrix} A_2 & B_2 \\ D_2 & A_2 \end{bmatrix} \times \ldots \begin{bmatrix} A_n & B_n \\ D_n & A_n \end{bmatrix} \quad \text{for a wall of n layers}$$

where

t_o, t_i = temperatures at surface o or i

q_o, q_i = heat flow at surface o or i

R_{so}, R_{si} = resistance of surface o or i

$A = \cosh\,(1 + j)\phi$

$B = \dfrac{R}{(1 + j)\phi} \cdot \sinh(1 + j)\phi$

$D = \dfrac{(1 + j)\phi}{R} \cdot \sinh(1 + j)\phi$

$\phi = \left(\dfrac{\omega b^2}{2\alpha}\right)^{0.5}$

$\omega = \dfrac{2 \cdot \pi}{24}$

$j = \sqrt{-1}$

b = thickness of layer (m)

$\alpha = \dfrac{\lambda}{\rho \cdot c}$ = thermal diffusivity

ρ = density (kg/m^3)

c = specific heat capacity (J/kgK)

λ = conductivity (W/mK)

Listing of the subroutine DECFAC, based on this and written by T. Yamaguchi, is reproduced below in facsimile form.

```
C************************* SUBROUTINE : DECFAC *************************
C
      SUBROUTINE DECFAC(IORO,NLAYER,COND,DENS,SPEC,THICK,RESIT,
     1                  RSI1,RSO1,
     2       UVALUE,  DECREM,TIMLAG,  ADMITT,TIMADM,  SURFAC,TIMSUR)
C
C     CALCULATION OF ADMITTANCE,DECREMENT FACTOR AND TIME LAG
C     OF A GIVEN MULTI-LAYER ELEMENT
C
C     METHOD  : MILBANK/JACOB/PIPES METHOD
C     PROGRAM : T.YAMAGUCHI, FEB-1981, DEPT.OF ARCH, UNIV. OF QLD
C
C***********************************************************************
C--INPUT
C     IORO    : 1 = INSIDE TO OUTSIDE, 0 = OUTSIDE TO INSIDE
C               DEFAULT: IORO=0
C     NLAYER  : THE NUMBER OF LAYERS OF ELEMENTS (MAX=10)
C               ADJUST "FOLLOWING DECLARATION" FOR OTHER NO.
      REAL COND(10),DENS(10),SPEC(10),THICK(10),RESIT(10)
      COMPLEX A(10),B(10),D(10)
C
C     (J = 1 TO NLAYER) THE ORDER OF DATA IS FROM "OUTSIDE" TO "INSIDE"
C     COND(J) : CONDUCTIVITY OF THE MATERIAL [W/M.C]
C     DENS(J) : DENSITY                      [KG/M3]
C     SPEC(J) : SPECIFIC HEAT                [J/KG.C] X1./3600[WH/KG.C]
C     THICK(J): THICKNESS                    [M]
C     RESIT(J): AIR (CAVITY) RESISTANCE       [M2.C/W]
C     RSO1,RSI1: OUTSIDE,INSIDE SURFACE RESISTANCE [M2.C/W]
C
C--OUTPUT
C     IORO    : IF ERROR OCCURS IN INPUT DATA,
C               RETURN -VE VALUES AS AN ERROR INDEX. IORO=0, IF NO ERRORS
C     UVALUE  : U-VALUE          [W/M2.C]
C     ADMITT  : ADMITTANCE       [W/M2.C]
C     TIMADM  : TIME-LEAD ::: ADMITTANCE [HR]
C     DECREM  : DECREMENT FACTOR  [-]
C     TIMLAG  : TIME LAG          [HOURS]
C     SURFAC  : SURFACE FACTOR [-]
C     TIMSUR  : TIME-LAG ::: SURFACE FACTOR [HR]
C
C-----------------------------------------------------------------------
C                INPUT DATA                   !  TREATMENT / ERROR
C--------------------------------------------
C     THICK  !  DENS   COND   SPEC  !  RESIT  !
C-----------------------------------------------------------------------
C       0    !   0      0      0    !    +    !  CAVITY
C       ?    !   ?      ?      ?    !    +    !  CAVITY
C       +    !   +      +      +    !    0    !  SOLID LAYER
C
C      -1    !  -2     -4     -8    !    .    !  ERROR (SUM OF -VE NO.)
C
C                                              -16 IF THICK > 6. [M]
C-----------------------------------------------------------------------
C   N.B. 0:ZERO OR -VE, +:+VE, ?:ARBITRARY
C        ERROR INDEX WILL BE RETURNED IN "IORO".
C***********************************************************************
C--FUNCTIONS : CSINH,CCOSH (HYPERBOLIC SINE & COSINE OF COMPLEX)
      COMPLEX CSINH,CCOSH
C
      COMPLEX UNIT,E,F,G,H,DENOMI,ADM,ZOI,ZWI,FOI
      COMPLEX ENEW,FNEW,GNEW,HNEW
      DATA PI/3.1415926/,UNIT/(1.,1.)/
      W=PI/12.  ! FREQUENCY OF HEAT INPUT [RADIAN/HOUR] (CYCLE=24 HOURS)
C***********************************************************************
C
C...DIRECTION SETTING [INSIDE TO OUTSIDE]
      J1=NLAYER
      J2=1
      J3=-1
      J4=NLAYER-1
      RSI=RSO1
      RSO=RSI1
      IF(IORO.EQ.1) GO TO 10
C...DIRECTION SETTING [OUTSIDE TO INSIDE]
      J1=1
      J2=NLAYER
      J3=1
      J4=2
      RSI=RSI1
      RSO=RSO1
C
10    TOTRES=RSO+RSI
C
      IERR=0
      DO 110 J=1,NLAYER,1
      IORO=0
      IF(RESIT(J).GT.0.) GO TO 50
      IF(THICK(J).LE.0.) IORO=IORO-1
      IF(DENS(J).LE.0.) IORO=IORO-2
      IF(COND(J).LE.0.) IORO=IORO-4
      IF(SPEC(J).LE.0.) IORO=IORO-8
      IF(THICK(J).GT.6.) IORO=-16
      IF(IORO.GE.0) GOTO 30
      IERR=IERR+1      ! COUNT NO. OF FAULTY LAYERS
      TYPE 20,J,IORO
20    FORMAT(' ??? ERROR IN INPUT DATA : LAYER=',I2,'  TYPE=',I4)
      IF(IORO.EQ.-16) RETURN  ! AVOID OVERFLOW IN CSINH,CCOSH.
30    CONTINUE
C--THERMAL DIFFUSIVITY [M2/H]
      THEDIF=COND(J)/DENS(J)/SPEC(J)*3600.     ! SPEC:J-->WH
C--THERMAL RESISTANCE  [M2.C/W]
      RESIST=THICK(J)/COND(J)
C--PHI
      P=SQRT(W*THICK(J)**2/THEDIF*0.5)
C--CALCULATE A,B,D VALUES FOR EACH LAYER
      A(J)=CCOSH(UNIT*P)
      B(J)=RESIST/(UNIT*P)*CSINH(UNIT*P)
      D(J)=(UNIT*P)/RESIST*CSINH(UNIT*P)
      GO TO 100
50    A(J)=1.
      B(J)=RESIT(J)
      D(J)=0.
      RESIST=RESIT(J)
100   CONTINUE
C--TOTAL RESISTANCE
      TOTRES=TOTRES + RESIST
110   CONTINUE
      IF(IERR.GT.0) RETURN
C
C...U-VALUE
      UVALUE=1./TOTRES
C
C--CALCULATE E,F,G,H FOR WHOLE ELEMENTS
      E=A(J1)
      F=B(J1)
      G=D(J1)
      H=A(J1)
      IF(NLAYER.LE.1) GO TO 200
C
      DO 150 J=J4,J2,J3
      ENEW=E*A(J) + F*D(J)
      FNEW=E*B(J) + F*A(J)
      GNEW=G*A(J) + H*D(J)
      HNEW=G*B(J) + H*A(J)
      E=ENEW
      F=FNEW
      G=GNEW
      H=HNEW
150   CONTINUE
C
200   DENOMI=E*RSI + RSI*RSO*G + H*RSO + F
      DENOMI=1./DENOMI
      ADM   =(G*RSI + H)*DENOMI
      ZOI   =1./ADM
      ZWI   =ZOI-RSO
C...ADMITTANCE
      ADMITT=CABS(ADM)
      TIMADM=TIMERD(ADM)/W
C...DECREMENT FACTOR
      DECREM=CABS(DENOMI)*TOTRES
      TIMLAG=TIMERD(DENOMI)/W
C...SURFACE FACTOR
      FOI=ZWI*ADM
      SURFAC=CABS(FOI)
      TIMSUR=TIMERD(FOI)/W
C
      RETURN
      END
C
C***********************************************************************
C
      FUNCTION TIMERD(CMPLEX)
C
C     CALCULATE ARC-TANGENT OF (S/R) IN RADIANS
C     WHERE R, S = REAL, IMAGINARY PARTS OF THE GIVEN COMPLEX VALUE.
C     CMPLEX = R + S * SQRT(-1.)
C
C***********************************************************************
      COMPLEX CMPLEX
      DATA PI/3.1415926/
C
      ATA=AIMAG(CMPLEX)/REAL(CMPLEX)
      ASA=AIMAG(CMPLEX)/CABS(CMPLEX)
      ASINE=ASIN(ASA)
      IF(ASA.GE.0..AND.ATA.GE.0.) TIME=ASINE
      IF(ASA.GE.0..AND.ATA.LT.0.) TIME=PI-ASINE
      IF(ASA.LT.0..AND.ATA.GE.0.) TIME=PI-ASINE
      IF(ASA.LT.0..AND.ATA.LT.0.) TIME=2.*PI+ASINE
      IF(TIME.GT.PI*2.) TIME=TIME-PI*2.
      TIME=2.*PI-TIME
      TIMERD =TIME
C
      RETURN
C***********************************************************************
      END
```

APPENDIX 4

STATISTICAL HANDLING OF TEMPERATURE DATA

The use of mean temperatures only can give quite misleading results. Some examination of the distribution (or variability) of temperatures is necessary.

The Bureau of Meteorology publishes, for a large number of locations in Australia the 14th and 86th percentile values of both the maxima and the minima for each month. (These values are chosen, as the 14th percentile would be the temperature not exceeded one day per week and the 86th would be that exceeded one day per week.)

What we need is the 14th and 86th percentiles of all temperatures, not only of the maxima and minima, as an indication of the overall frequency distribution of temperatures. From the published values the standard deviation of maxima and minima can be found:

$$sd(max) = \frac{T_{max86} - T_{max14}}{2.16}$$

$$sd(min) = \frac{T_{min86} - T_{min14}}{2.16}$$

The standard deviation of the overall temperature distribution will be:

$$sd = 0.5\ [sd(max)^2 + sd(min)^2 + 2 \cdot r \cdot sd(max) \cdot sd(min)]^{0.5}$$

Walsh and Spencer (16) concluded that a value of 0.4 for the correlation coefficient 'r' (between daily minima and maxima) gives a result of acceptable accuracy for all Australian locations.

From this the two values wanted can be calculated as

$$T_{14} = \overline{T} - 1.08\ sd$$

$$T_{86} = \overline{T} + 1.08\ sd$$

At present these values must be pre-calculated and are required as input in the LOCAT sub-program. At a later date a small subroutine may be included to perform this calculation.

APPENDIX 5

ALGORITHMS FOR IRRADIANCE CALCULATIONS

A. If hourly values of beam and diffuse irradiance, G_b and G_d, measured on a horizontal plane are given, as mean values for the month:

(1) determine declination, using NDY (number of day of year) for the 15th of each month

$$DEC = 23.45 \sin \left[\frac{360}{365} (NDY + 284)\right]$$

(2) find sunrise time (SRT) and sunset time (SST)

$$SRT = 12 - \frac{\cos^{-1}(-\tan LAT \cdot \tan DEC)}{15}$$

$$SST = 24 - SRT$$

where LAT = geographical latitude (south -ve)

(3) find hour angle for each hour

$HRA = 15\ (HR-12)$ thus am -ve, pm +ve. HR = hour, 24-hour clock

but, if sun rises during the hour, then $HR = \frac{HR + SRT}{2}$

or if sun sets during the hour, then $HR = \frac{HR-1+SST}{2}$

(4) find solar position angles, altitude (ALT) and azimuth (AZI)

$$ALT = \sin^{-1}(\sin LAT \cdot \sin DEC + \cos LAT \cdot \cos DEC \cdot \cos HRA)$$

$$AZI = \cos^{-1}\left(\frac{\cos LAT \cdot \sin DEC - \sin LAT \cdot \cos DEC \cdot \cos HRA}{\cos ALT}\right)$$

for pm hours $AZI = 360 - AZI$

(5) find angle of incidence, but first the horizontal shadow angle for surface with orientation ORI

$HSA = AZI - ORI$ check: if $90 < |HSA| < 270$ then $HSA = 0$
if $HSA > 270$ then $HSA = HSA - 360$
if $HSA < -270$ then $HSA = HSA + 360$

$$INC = \cos^{-1}(\sin LAT \cdot \cos TIL + \cos HSA \cdot \cos ALT \cdot \sin TIL)$$

if $INC > 90$ then $INC = 90$

where TIL is tilt angle from horizontal, +ve if south facing, -ve if north

(6) the irradiance on surface 's' will be

$$G_s = G_b \frac{\cos INC}{\sin ALT} + G_d \frac{1+\cos TIL}{2} + \rho(G_d+G_b) \frac{1-\cos TIL}{2}$$

then the 24-hour mean irradiance will be

$$\bar{G}_s = \frac{1}{24} \sum_1^{24} G_s$$

where ρ = reflectance of foreground

APPENDIX 5 (continued)

B. If only the daily total irradiation (H) is given, then first the beam and diffuse components must be estimated.

(1) find irradiation on the same plane in the absence of any atmosphere

$$H_o = 1353\ [1 + 0.033 \cos(\frac{360}{365})\ NDY)]\ COE\ \frac{24}{\pi}$$

where the coefficient is

$$COE = \cos LAT \cdot \cos DEC \cdot \sin SHA + \frac{\pi}{180}\ SHA \cdot \sin LAT \cdot \sin DEC$$

and the sunrise hour angle: $SHA = \cos^{-1}(-\tan LAT \cdot \tan DEC)$

(2) find 'cloudiness index': $K = \frac{H}{H_o}$ and with this the

'diffuse ratio':

$$R = 1.3903 - 4.0273\ K + 5.5315\ K^2 - 3.108\ K^3$$

(3) then the diffuse irradiation is $H_d = R \cdot H$

and the beam irradiation: $H_b = (1-R) \cdot H = H - H_d$

(4) Both beam and diffuse components will be divided into hourly irradiation (Wh/m^2) or hourly mean irradiance (W/m^2) values, which are numerically the same, using the proportions given by Liu and Jordan (17).

REFERENCES

1. H.J. Cowan, editorial in Arch. Sc. Rev. Nov. (1959).

2. Petherbridge, P. (1974) Limiting temperatures in naturally ventilated buildings in warm climates, BRE.CP 7/74.

3. N.O. Milbank and J. Harrington-Lynn, Thermal response and the admittance procedure, The Bldg. Serv. Eng. (J.IHVE) 42, 38-51 (1974).

4. P. Petherbridge, private communication (1980).

5. E. Danter, Periodic heat flow characteristics of simple walls and roofs, J.IHVE 28, 136-145 (1960).

6. C.O. Mackey and L.T. Wright, Periodic heat flow - homogeneous walls or roofs, Heating, Piping, Air. cond., 546-555 (1944) also ASHVE Trans. 50, 293

7. C.O. Mackey and L.T. Wright, Periodic heat flow - composite walls or roofs, ASHVE Trans. 52, 288-296.

8. Ritson, P.R. (1982) Computer simulation of thermal performance for building design, M.Arch. thesis, Uni. of Qld.

9. Mitchell, W.J. (1977) Computer-aided architectural design, van Nostrand Reinhold, New York.

10. Regulation establishing energy conservation standards for new residential and non-residential buildings, CON-45A: 01, Calif. Energy Resources, Conservation and Development Commission, 1976/1978.

11. Auliciems, A. (1981) Global differences in indoor thermal requirements, ANZAAS Conf., Brisbane.

12. Woolard, D.S. (1979) Thermal habitability of shelters in the Solomon Islands, Ph.D. thesis, Uni. of Qld.

13. Turner, S.O. (1981) A thermal performance assessment of architect-designed houses in Brisbane, B.Arch. thesis, Uni. of Qld.

14. Jakob, M. (1949) Heat transfer, Vol. 1, John Wiley.

15. L.A. Pipes, Matrix analysis of heat transfer problems, J. Franklin Inst., 623, 195-206 (1957).

16. Walsh, P.J. and Spencer, J.W., (1980) Heating degree-days for Australian localities, CSIRO Div. Bldg. Res. Tech. Paper 35.

17. B.Y.H. Liu and R.C. Jordan, The interrelationship and characteristic distribution of direct, diffuse and total solar radiation, Solar Energy 4(3), 1-19, July (1960).

An Evaluation of Thermal Performance Computer Programs

T.J. Williamson
Department of Architecture,
University of Adelaide

ABSTRACT

A number of computer programs are available which can be used to estimate the thermal performance of buildings. This paper reports on the development of a methodology to evaluate such programs in order to provide comprehensive information on their capabilities to potential users. Some pitfalls in using these programs are highlighted.

INTRODUCTION

The key to improving the thermal performance of buildings and the efficient use of energy for heating and cooling is the ability to accurately predict at the pre-construction stage, the performance of the building envelope and plant, and to assess the consequences of operating the building in different ways. This type of information is required at many levels of the building industry; from the individual designer wishing to test the performance of a particular building, to government policy-makers wishing to evaluate the likely effectiveness of proposed incentive schemes or mandatory requirements aimed at reducing energy consumption. Buildings constructed without such predictions being undertaken, or schemes introduced without thorough evaluation, must be considered simply as experiments and the outcome uncertain. Such a hit and miss approach to the problems of improving the thermal performance of buildings is likely to prove costly, wasteful of valuable resources, and could impose unnecessary thermal discomfort on the occupants.

The thermal behaviour of a building is a complex interaction between its configuration, the climate, the site, the properties of the building elements and finally, the occupant behaviour. Most factors are continuously varying. Manual evaluation techniques have not yet been developed which satisfactorily take into account all significant variables. Relatively simple computation techniques are available for estimating the required heating and cooling plant capacities and energy consumptions. The accuracy of these calculations however is questionable and their flexibility limited. Perhaps the most serious deficiency is that simple techniques do not exist for estimating the thermal performance of buildings without air-conditioning which is the most usual case found in the majority of buildings in Australia.

Computer programs offer the only practical method of generating the required information. Several programs have been developed both in Australia and overseas which purport to be suitable for this application. This paper outlines a project conducted under the auspices of the Australian Housing Research Council to evaluate such computer programs in the Australian context, and summarises some important findings.

AIM AND STRUCTURE OF PROJECT

The overall aim of the project was to develop a methodology by which thermal performance computer programs (TPCPs) could be evaluated and to apply the methodology to selected programs. It was realised that to be comprehensive the evaluation methodology must meet the needs of a variety of anticipated program users. The information required by potential users will vary and will depend upon their individual requirements. Three general areas of likely questions regarding TPCPs were identified. These are:

(a) The Applicability of the Program
The relevance of the program to various users such as architects, owners of existing buildings, public housing authorities, research workers, etc. The range of thermal performance problems which can be solved by the program. The types of buildings which may be analysed.

(b) The Operational Characteristics of the Program
Program language, computer requirements, data necessary, ease of operation, input, operating cost, acquisition cost.

(c) The Program Precision
Accuracy in predicting the thermal performance of dwellings. Thermal performance is defined broadly as the reaction of the building to all external and internal factors. This takes into account both unconditioned and conditioned dwellings, and the range of climates found within Australia. The relationship of the program description of thermal performance (e.g. temperature) and occupant thermal comfort must also be considered.

Accuracy in predicting required plant capacity and energy consumption. It is necessary to differentiate between calculated energy required to condition a space and total energy consumption considering operational and equipment characteristics.

When designing the general form of the evaluation methodology to cover these general questions, two separate components were identified as being necessary. First, a report should be prepared by program owners covering points (a) and (b) above, and secondly, an assessment of program accuracy should be conducted by an independent body.

To ensure program owners provided sufficient information for potential users to make meaningful comparisons between programs, a standard reporting format was developed. This required owners to provide the following information.

General Details of Program

- purpose of the program
- program author, when developed
- program language
- mathematical basis
  - heat flow through opaque elements
  - calculation of heating and cooling loads
  - thermostating procedures
  - include reference papers
- nature and significance of results
- accessibility of program to potential users
  - where program is available
  - cost of purchasing program
- transferability of program to different computers
  - size of computer
- program flexibility, variables which can be considered, particularly user characteristics
  - zoning
  - plant operation
  - window controls, e.g. shading, opening and closing
  - occupant loads, e.g. lighting, cooking
- existence, comprehensiveness and simplicity of operating manual
- any editions of program for special users
- optional sub-routine available

Input

- input data required
- availability of required data
  - climatic
  - building
  - user characteristics
- for a typical house, form, time and cost of data preparation
- skills required to assemble data

Operation

- mode of operation
  - batch
  - interactive
- skills required to run simulation

- for a typical house, computer costs for simulation run
  - specify length of run, e.g. day, annual
  - specify type of computer
  - internal user rate, external user rate

Output
- form of output, options available
- ease and cost of evaluating design and behavioural variations
- skills required to interpret output information

Applications
- evaluations of program carried out to date
- applications of program to date
- potential applications of program
- planned or feasible developments which would extend the applications of the program

It was decided that the most realistic assessment of program accuracy could best be achieved by comparing program predicted values of thermal performance with monitored values. A full assessment of program capabilities would require sets of data to be compiled representing a variety of:

- building types and sites
- climate
- building operation modes
- plant type and operation

To ensure credibility for this component of the evaluation program owners were not to have access to monitored data until after a simulation run had been completed.

EVALUATION OF PROGRAMS

Choice of Programs

The proposed methodology of program evaluation was tested by applying it to four TPCPs. Two basic criteria were used in choosing these initial programs and followed principally from the requirements of public housing authorities.

(i) The program purported to contain a substantial component directed at modelling the thermal performance of a building, particularly in an unconditioned state. Programs designed primarily to estimate air-conditioning loads for buildings were excluded as these were the subject of a special study conducted by The Association of Computer Aided Design (1). Programs which simulate the performance of building plant, e.g. solar air-conditioning, but which do not attempt, other than by very simple models, to predict the thermal behaviour of the complete building were similarly excluded.

(ii) The program was extensively used in Australia and experienced operators were available. Various imported programs were discovered on files of commercial computing organisations, but generally no experience of their operation existed in Australia.

The four programs chosen for evaluation are well known in Australia. Three were developed in Australia and one imported from the United States of America but modified for Australian conditions. The programs will be identified simply as A, B, C and D *.

Program Reports

The owners of each program were invited to submit a report according to the format outlined above. The responses showed variation, the significance of which would bewilder a potential user not familiar with the terminology. Table 1 gives an extract from each program report of two items - program purpose and program flexibility. The primary purpose of these reports is to highlight gross differences between program capabilities and user requirements. The significance of the subtle differences between programs, such as the method of handling shading, cannot be ascertained from such reports.

TABLE 1 Summary of Program Purpose and Flexibility

Pro-gram	Purpose	Flexibility
A	A program for the calculation of the thermal performance of buildings. For a given building in a given location the program calculates indoor environmental temperatures and sensible heating/cooling loads over as long a time period as required using real climatic data if necessary.	Up to ten zones can be considered for one building. Plant operation may be specified in terms of a capacity, thermostat setting and thermostat differential. The thermostat setting may be varied during the 24 hour day. Heating requirements may be calculated for ten zones in one building, cooling requirements in only one. Horizontal shading of vertical building surfaces by eaves is a standard feature. Shading by trees, other buildings is not. Only vertical building surfaces oriented mutually at right angles (e.g. N-E-S-W) may be considered. Only a horizontal roof may be considered. The rate of infiltration/ventilation is constant.

* The conditions of confidentiality under which the project was conducted preclude explicit identification of the program at this time.

TABLE 1 Summary of Program Purpose and Flexibility (cont.)

Program	Purpose	Flexibility
B	This program is directed to the prediction of thermal performance of dwellings, although it may be used for other classes of buildings. Using actual weather data or 'design' data, it predicts hourly environmental temperatures, sensible heating/cooling loads and the total energy added to or extracted from spaces for any nominated period - 3 days to years. The program is essentially design/user orientated.	Two habitable zones may be considered. The program estimates hourly heating and/or cooling to zones to meet specified internal conditions for any period. If limited capacity plant is specified, hourly estimates of temperatures attained are calculated, or if no heating or cooling, the hourly temperatures attained. Thermostat temperatures may be specified with a time on and off and a plus and/or minus temperature swing. The program is limited to ten each of vertical surfaces, horizontal surfaces and inclined surfaces. Similarly the total number of facade orientations is limited to ten. In use, different surfaces of the same orientation associated with the same zone may be combined. All shading configurations may be considered. Infiltration/ventilation is calculated as a function of wind speed.
C	This program can be applied to most types of buildings; glasshouses should be treated with caution. The program calculates the sensible and latent air conditioning loads (heating or cooling) for a specified indoor air temperature and humidity within a building (LOAD), or alternatively, the indoor air conditions for a specified heating or cooling effect (TEMP).	Zoning is at the user's discretion and does not necessarily have to be associated with building features or facades. Twenty zones are permitted. In the LOAD mode, the indoor temperatures specified by the user are assumed held constant by the plant. The plant is given one hour at start-up to bring the indoor temperature up or down to the control level. In the TEMP mode the plant is controlled from the thermostat. Hours of plant operation may be user specified in both cases. Window controls: the user may choose a shading coefficient appropriate to sun on the window and one for when the direct sun has moved off a window. These are then applied automatically according to the orientation of any particular window. Variable ventilation can be accommodated (2 levels) according to the difference between room temperature and ambient (cooling only). Occupancy, lighting and internal loads are all modified. The activity rate of occupants can be selected from one of four levels. Lighting can be split into radiative and convective fractions. Both of these are associated with on/off times, user specified. Additionally, a fixed load can be input; and a 12 time-step variable internal load profile can be called up.
D	This program is a transient simulation package for calculating the behaviour of buildings and building components with particular emphasis on the use of solar based energy systems. The behaviour of the building and its sub-systems can be studied over a period of time at specified increments of time. Although much of the program is concerned with energy systems, the behaviour of the building as a whole may also be examined.	The flexibility of this package is one of its greatest strengths. The user is free to define his system in any way he wishes, using the available components or specially written component sub-routines. In most cases, several components of the same type can be included in the system. This flexibility allows the user to define unique simulation systems to his own requirements. The interconnections between components is also left to the user.

Comparison of Predicted and Monitored Thermal Performance

The accuracy of programs was assessed by comparing predicted values of internal temperatures with monitored internal conditions. Collection of long term energy consumption data was not possible, therefore, energy consumption predictions were compared only between programs, simulating the same building with constant climatic factors and operating conditions.

In all, six dwellings in different locations throughout Australia were monitored. These provided a variety, but by no means a full range, of climatic conditions, and construction forms. The dwellings were unoccupied during the monitoring period, but in-use conditions were simulated by operating heating appliances, opening windows and adjusting window controls. Each program owner was given the same information consisting of:

- dwelling plan and elevations
- site plan
- details of construction and site conditions
- hourly values of site measured external climate
  Table 2 shows the form and content of this data
- measured infiltration rates
  Infiltration rates measured by the gas concentration decay technique and correlated with wind speed.
  This was expressed in the form:
  $$ACH = A + B.V$$
  where ACH - ventilation rate, air changes per hour
  V - wind speed
  A and B - constants for particular dwelling
- details of plant and window operation

The results of recorded internal conditions were not made available to program owners until after an initial simulation run had been submitted. At that stage each owner was given the opportunity to amend their assumed input data or program and submit a further simulation. The results of computer predictions of performance for only two of the monitored dwellings will be presented here. Details of these dwellings, one in Melbourne (Broadmeadows) and the other in Brisbane (Goodna) are given in Appendix A.

Also shown for each location are, first, the external climate comprising measured dry bulb temperature and solar radiation, and secondly, internal environmental temperatures recorded at several points in the living zone of the house. These graphs represent one week of a total of approximately four consequetive weeks' monitoring at each location.

The internal environmental temperatures were calculated from measurements of dry bulb temperature and globe temperature according to the following relationship:

TABLE 2 External Climate Data

BR78120902	202	174	2 2 020	0 0
~~BR78120903~~	~~192~~	~~168~~	~~1 2 030~~	~~0 0~~
BR78120904	186.	165	1 1 030	0 0
~~BR78120905~~	~~184~~	~~162~~	~~1 1 030~~	~~2 1~~
BR78120906	188	162	1 1 030	100 60
~~BR78120907~~	~~203~~	~~175~~	~~1 2 030~~	~~306 73~~
BR78120908	232	187	3 3 040	544 86
~~BR78120909~~	~~254~~	~~194~~	~~3 31560~~	~~709 85~~
BR78120910	274	207	3 41560	862279
~~BR78120911~~	~~299~~	~~213~~	~~5 41660~~	~~867346~~
BR78120912	308	223	4 5 180	892428
~~BR78120913~~	~~306~~	~~224~~	~~5 5 180~~	~~847499~~
BR78120914	291	218	4 41680	518465
~~BR78120915~~	~~296~~	~~221~~	~~6 51680~~	~~484419~~
BR78120916	284	213	6 51680	273270
~~BR78120917~~	~~265~~	~~203~~	~~4 41580~~	~~145150~~
BR78120918	265	198	2 21580	45 38
~~BR78120919~~	~~248~~	~~196~~	~~2 31580~~	~~3 2~~
BR78120920	241	194	2 31340	0 0
~~BR78120921~~	~~235~~	~~190~~	~~2 21330~~	~~0 0~~
BR78120922	235	191	2 21330	0 0
~~BR78120923~~	~~234~~	~~190~~	~~1 11340~~	~~0 0~~
BR78121000	225	188	1 11550	0 0
~~BR78121001~~	~~217~~	~~185~~	~~1 11540~~	~~0 0~~
BR78121002	215	180	1 1 030	0 0

cols	1- 2	Location (e.g. BR = Brisbane)
	3- 4	Year (e.g. 79)
	5- 6	Month (01-12)
	7- 8	Day (1-31)
	9-10	Hour (0-23) apparent solar time, 0 = midnight
	11-14	External Dry Bulb Temperature (°C x 10)
	15-18	External Wet Bulb Temperature (°C x 10)
	19-20	Blank
	21-22	Wind Speed 1 (knots)
	23-24	Wind Speed 2 (knots)
	25-26	Wind Direction (0-16, 0 = calm, 1 = NNE)
	27	Cloud Cover (oktas, 0-8)
	28	Blank
	29-32	Global Irradiance on Horizontal Plane (W/m^2)
	33-35	Diffuse Irradiance on Horizontal Plane (W/m^2)
	35-40	Blank

$$t_e = \frac{6}{5} t_g - \frac{1}{5} t_a$$

where t_e - environmental temperature

t_g - globe temperature

t_a - air dry bulb temperature

Derivation of this relationship is given in Appendix B.

Environmental temperature was used as an indicator of thermal performance because it gives a better indication of thermal comfort than just dry bulb temperature.

Figs. 1 and 2 show examples of various computer predictions for the Melbourne and Brisbane locations, compared with monitored mean internal environmental temperature.

The significance of these results must be left to each potential user to assess. This however, may not be easy. If the program assessment result shows a difference between predicted and monitored values one can never be certain what is causing the error. Any one of a number of faults may contribute to the discrepancy. The most significant sources of error are however likely to be:

- inappropriate simplifying assumptions, e.g. constant ventilation, constant shading factor
- incorrect data input

Likewise a precise simulation of one building is not proof of the accuracy of the program. Only consistently good predictions over a number of building types and situations can lead to confidence in the results produced by a program.

Energy Consumption Calculations and Loads

The owners of two programs, A and B, offered an estimate of the likely annual energy consumption for heating and cooling requirements and corresponding maximum plant loads for the two dwellings. Both programs used identical climatic data and operating conditions. The results are given in Table 3.

TABLE 3 Estimated Energy, Annual Consumptions (Heating and Cooling) and Maximum Loads

	Dwelling Location			
Program	Broadmeadows		Goodna	
	Winter (heating) May - Nov.	Summer (cooling) Dec. - Apr.	Winter (heating) May - Sept.	Summer (cooling) Oct. - Apr.
A	15.6 G.J. $6 < L$ max. < 7 kW	4.4 G.J. $6 < L$ max. < 7 kW	4.0 G.J. $5 < L$ max. < 6 kW	13.8 G.J. $6 < L$ max. < 7 kW
B	15.7 G.J. $5 < L$ max. < 6 kW	2.1 G.J. $4 < L$ max. < 5 kW	8.2 G.J. $9 < L$ max. < 10 kW	12.1 G.J. $5 < L$ max. < 6 kW

Again the reason for the variations can never be known. Until program predictions can be validated by comparing them with measured energy consumption figures care must be exercised when using the results of any program.

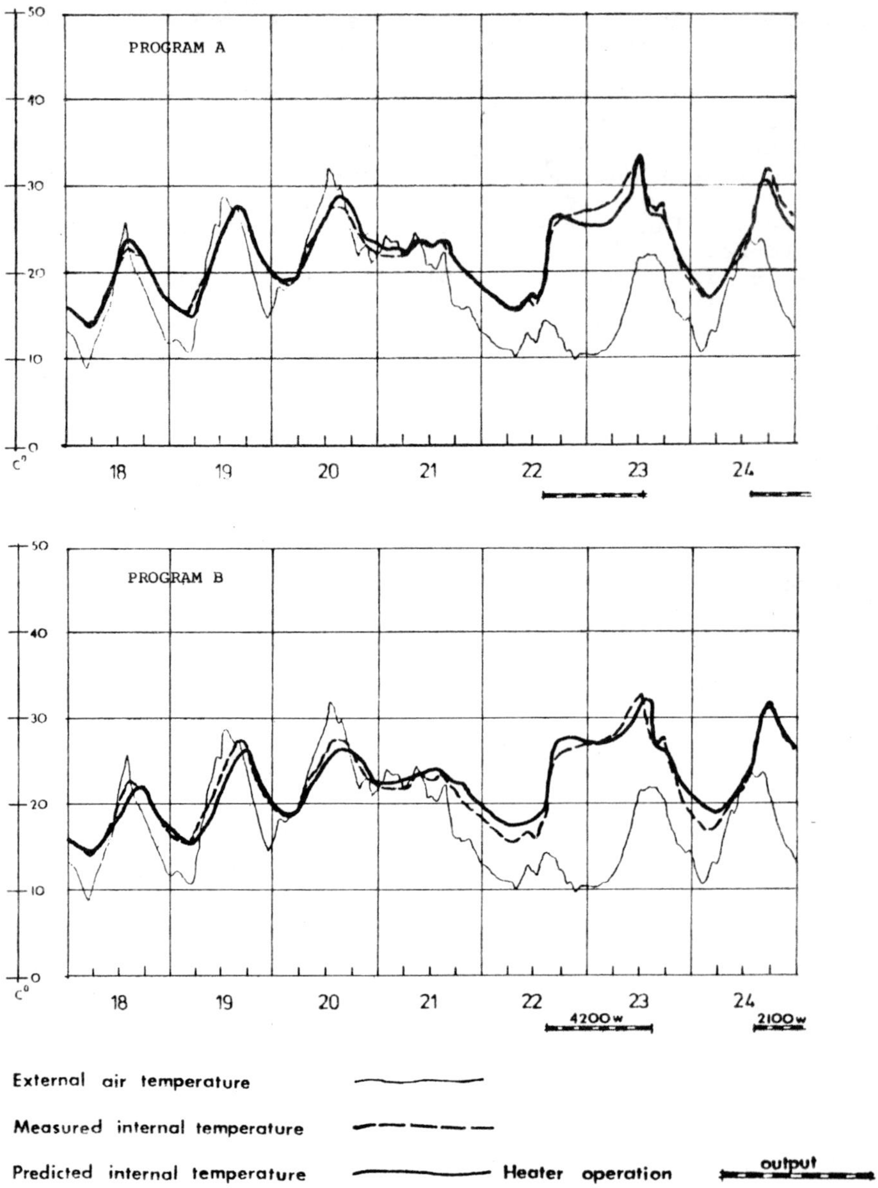

Fig 1 Comparison of Predicted and Monitored Internal Temperatures, Broadmeadows House

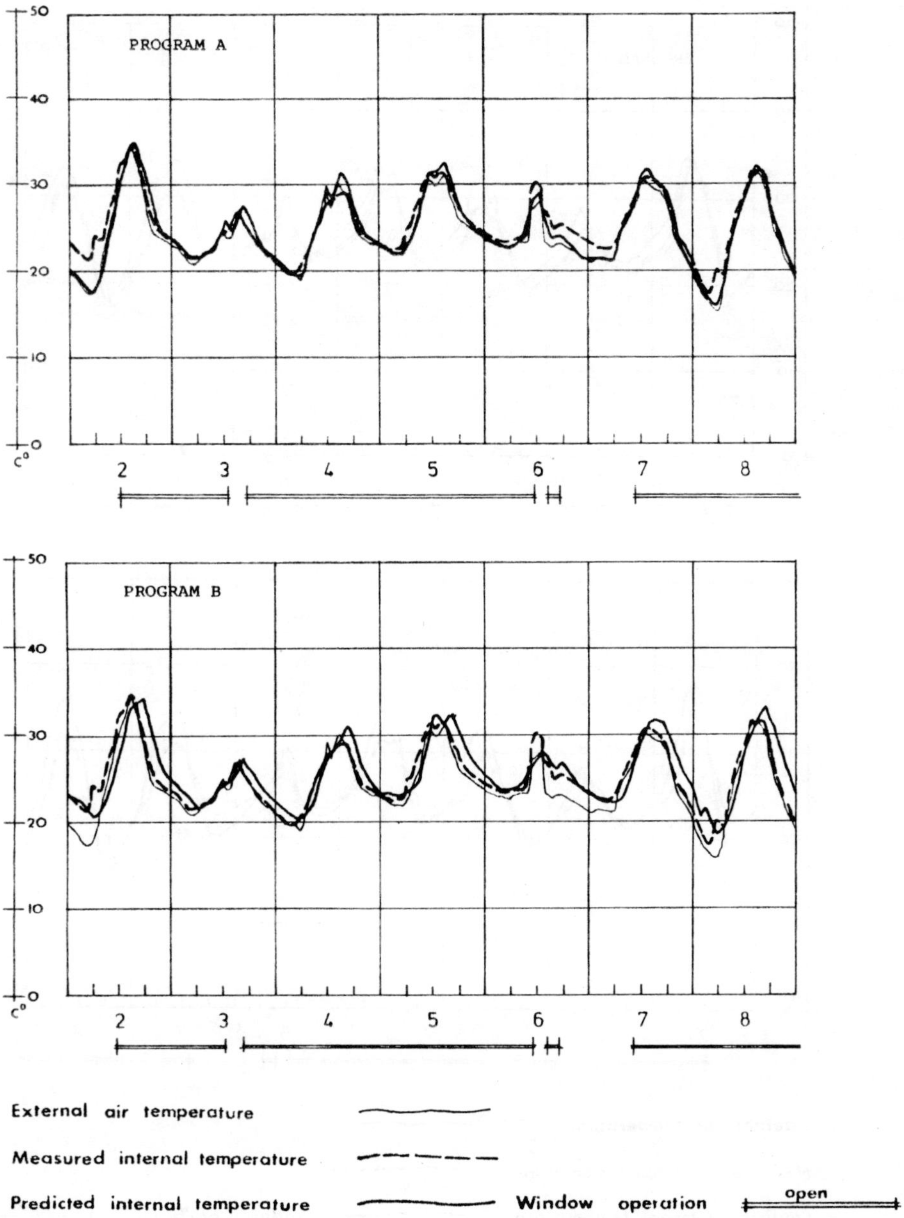

Fig. 2 Comparison of Predicted and Monitored Internal Temperatures, Goodna House

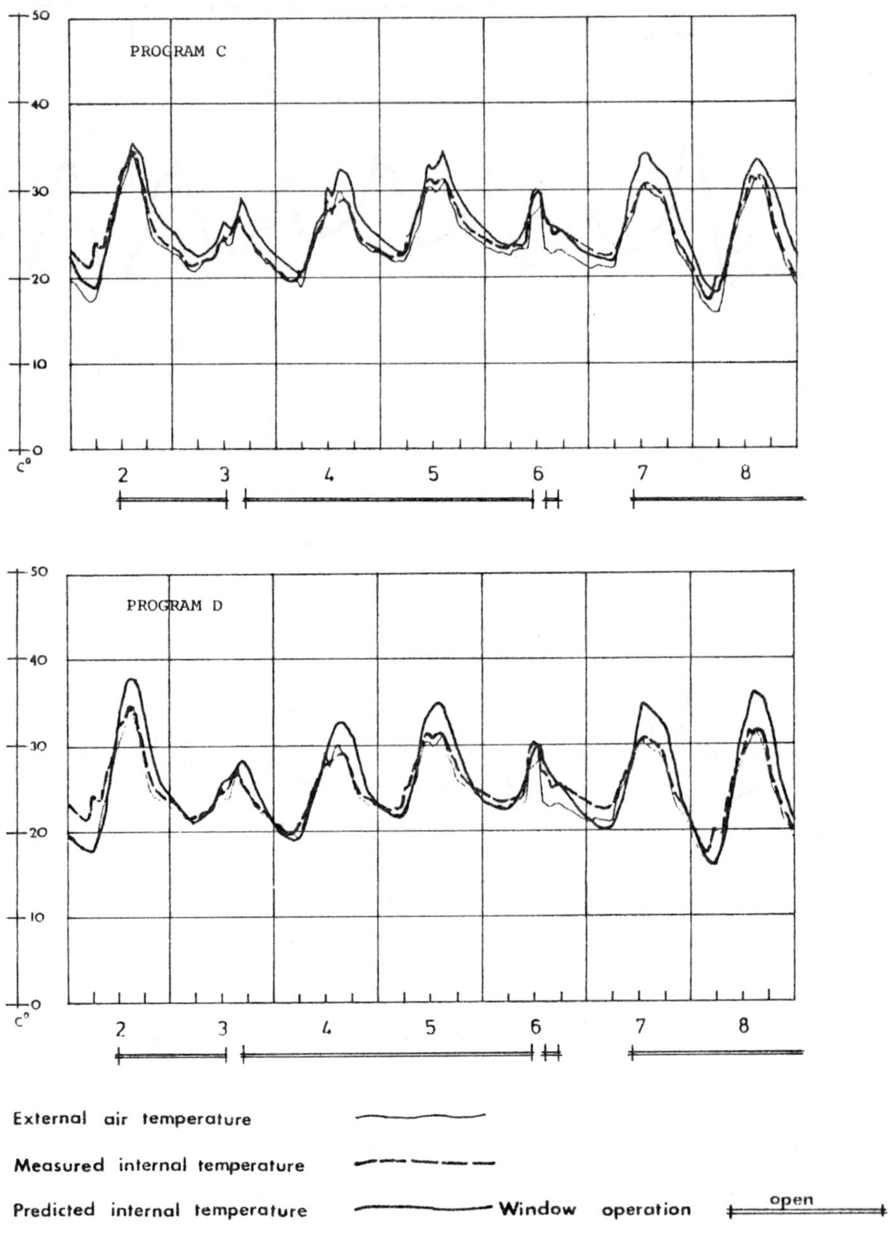

Fig. 2 Comparison of predicted and monitored internal temperatures, Goodna House (cont.)

THE RELEVANCE OF THERMAL PERFORMANCE COMPUTER PROGRAMS

The TPCP simulation of a building is by no means an exact science. The truth of this statement is all too obvious when you compare the results presented above. Many of the variables inherent in the problem can never by accurately known, and assumptions and guesses must be made. Nor is a computer simulation costless, or even rapid in a total sense. Therefore, in using any program one must be aware of trade-off between the cost and inconvenience, and the expected benefits expressed in terms of the information generated by the program. The nature of this trade-off will depend on the requirements of each potential user, the operational characteristics of the program and the confidence one has in the results.

The following sections detail some of the points which must be kept in mind when using or considering using TPCPs, together with some recommendations flowing from the project.

DATA REQUIREMENTS

The input data required by most TPCPs relates to three factors:

- external climate
- building envelope, ventilation and shading
- occupancy and operational patterns

External Climate Data

The external climatic data used by a program may be of two basic types:

(i) actual measured values
or (ii) internally generated values

All four programs used in this study employed a version which required measured coincident hourly values of the climatic variables:

- dry bulb temperature
- wet bulb temperature
- direct solar radiation
- diffuse solar radiation
- wind speed

Although actual weather data should theoretically produce the most realistic results a serious limitation exists because compiled data is available for only 16 locations in Australia, as indicated in Fig. 3, and data does not exist for two capital cities, Sydney and Adelaide.

Some programs have versions which generate clear sky values for solar radiation and a diurnal sinuosoidal temperature variation given the mean temperature and maximum temperature swing. These versions obviously are useless for relating to expected actual internal conditions or energy consumptions, but may be useful for ranking alternative design solutions.

Fig. 3 Locations of Compiled Climatic Data

Building Envelope, Ventilation and Shading Data

The thermal properties of the building envelope must be used as input data to all thermal performance programs. The method of data input relates to the heat flow algorithm used in the program. Standard values of properties are generally recommended for use, but one can never be certain that the properties of the materials used in the building construction correspond to these standard values. In fact, it is most unlikely.

A good example is the assumptions which must be made regarding ground conditions. Errors in the thermal resistance and capacitance of the soil can be particularly significant if slab-on-ground construction is used. An experienced program operator can be expected to make consistent guesses and therefore enable a true comparison of results even if the 'correct' answer is not strictly obtained.

The lack of precision in both knowing, and being able to model the complex variations of air infiltration and ventilation leads to a lack of confidence in any results obtained. Even slight errors in assumptions concerning this variable can be significant. The inaccuracies displayed in attempts to match monitored with predicted conditions for the test programs can be attributed substantially to this factor.

Accurate modelling of complex shading conditions generally entails tedious and time consuming data preparation and input. Most programs simplify this requirement by using a constant shading coefficient or similar.

Occupancy and Operational Patterns

The living pattern of the occupants and their mode of operating the building are ultimately the most significant factors in determining thermal behaviour and energy consumption. This factor is also the one we can know least about, especially as the final occupant of most buildings is not known at the time of design.

In the design situation therefore a standard occupancy must be assumed and all proposals assessed on this basis. In other situations various occupancy patterns can be used in order to assess the sensitivity of this factor.

One must always remember that TPCPs are concerned primarily with evaluating the 'physical' components of a building, such as envelope design, siting, etc. The basic philosophy in using the programs is to optimise these factors, given all the various constraints and objectives. In so doing the occupants are provided with the best opportunity for reducing energy consumption and for achieving a satisfactory thermal environment. It is up to the occupants to fully utilise the possibilities of the building.

PROGRAM USER

Care must always be exercised when interpreting the results of a computer simulation. The significance of the output must be fully understood, and for the relatively large and complex programs considered in the study this requires a degree of experience and knowledge of the operations of the program. The influence of program user on the results obtained from TPCPs has been highlighted by exercises involving a number of users analysing the same building with the same program (2). Due to differing assumptions regarding the building envelope properties, ventilation rates, etc., and errors with the input data, variations in the results were shown often to exceed 100%.

INTERNAL TEMPERATURE AND ENERGY CONSUMPTION PREDICTIONS

As stated above, the reasons for program errors can never be fully isolated. If program simulations are to be relied upon as the basis for design decisions then repeated accurate prediction of monitored values is the only way of being satisfied that the results are likely to be correct. Program owners should be required to provide a statement of probable accuracy based upon such comparisons.

Each of the programs used in the study claims to have the capacity to calculate the energy required to condition a space. The relevance of the figures produced must be understood. To be equated with actual energy consumption, the efficiencies and characteristics of heating and cooling plant must be taken into account, but none of the programs takes this into account. Nor do these programs model the interaction between building behaviour and plant operation, other than in the simplest fashion. It is well known however, that equipment/building operating characteristics can have a significant effect on actual energy consumption.

THERMAL COMFORT

Two of the programs included in this study produce output as environmental temperature, while the other two provide dry bulb temperatures.

Neither environmental or dry bulb temperatures correlate in all circumstances to the sensations of thermal comfort. In neither case are the effects of humidity and air movement taken into account. The estimation of a comfort index or similar must be calculated by a separate procedure, using the output from the program and combining the factors of wet bulb temperature or humidity and air movement. Although a model for estimating wet bulb temperature could be developed little is known about the real nature of air movement within a building.

No program can ever give all the answers and a program user is always required to make judgements about the overall thermal performance based upon data produced by the program.

A necessary first step in exercising this judgement is to derive a set of criteria to describe an acceptable level of thermal performance. The task of the program user is then to assess the building behaviour against these criteria and to determine the acceptability or otherwise of the design proposal. As an example, the performance criteria could be based upon the frequency of occurrence of high and low temperatures, e.g. high temperature performance in living areas could be stated in the form:

- 28°C should not be exceeded on more than 12% of the time

If the criteria were set in this form then programs which output a temperature/time frequency analysis will aid the program user to assess the results of a simulation more quickly.

SUMMARY

Computer techniques give the potential for modelling extremely complex phenomena, but if they are applied without sound judgement they can be worse than useless, time consuming and convey an impression of knowledge that does not exist. Part of the judgement required about a program relates to its accuracy in predicting actual conditions within a building. If one can be confident that energy consumption estimates, for example, are correct, the results can be used to undertake realistic economic assessment of alternatives. If the predictions are not entirely accurate the use of energy consumption estimates must be limited to ranking alternatives but without ever knowing the 'correct' answer. As there are few detailed accounts of computer predictions being correlated with monitored data over extended periods of time, one must assume that the majority of programs are intended only for this ranking function. Unfortunately this is not always the case and far too often more reliance is placed upon the results than would appear to be justified.

Finally, a legal implication of TPCP use should be considered. The provisions of the various Trade Practices Acts and Consumer Protection Acts may preclude the publication of the results from computer studies unless the program has been thoroughly validated.

The obvious conclusions from this study point to an urgent need to formulate a standard evaluation methodology to which all TPCPs may be subjected in order to verify their capabilities and validate their accuracy.

ACKNOWLEDGEMENTS

The contributions of the anonymous program owners must be recognised. Without their interest and assistance this work would not have been possible. The Australian Housing Research Council who supported the work financially is also gratefully acknowledged.

REFERENCES

(1) The Association for Computer Aided Design. (1977) Air Conditioning Load Programs, ACADS, Report No. 7. Melbourne.

(2) Spielvogel, L.G. "Comparison of energy analysis computer programs." ASHRAE Journal, 14(1), 45-47 (1978).

APPENDIX A

Details of Monitored Houses, Broadmeadows and Goodna

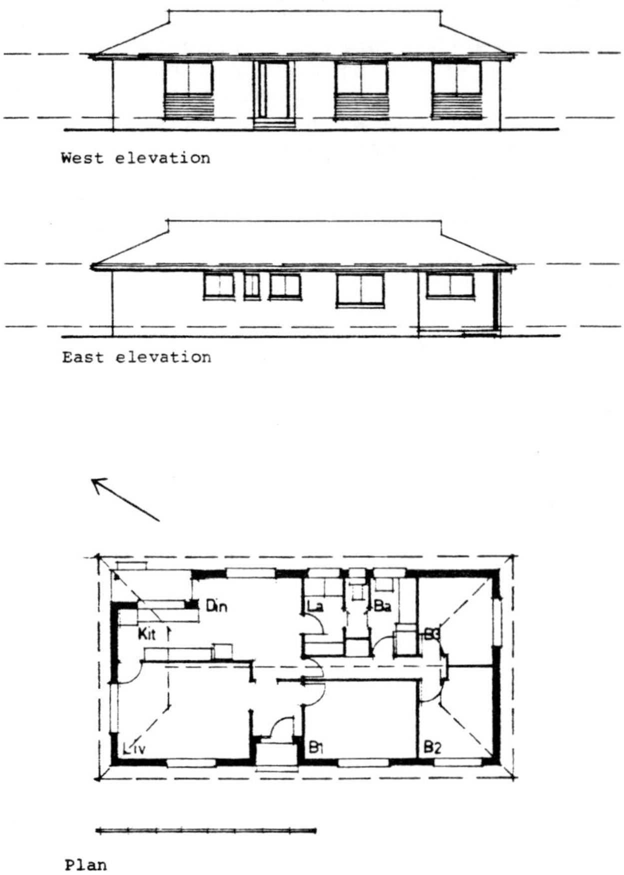

Construction - Brick veneer, concrete slab-on-ground, plasterboard ceiling with 75mm fibreglass insulation, concrete tile roof

Fig. A1.1 House Monitored at Broadmeadows, Melbourne

APPENDIX A (cont.)

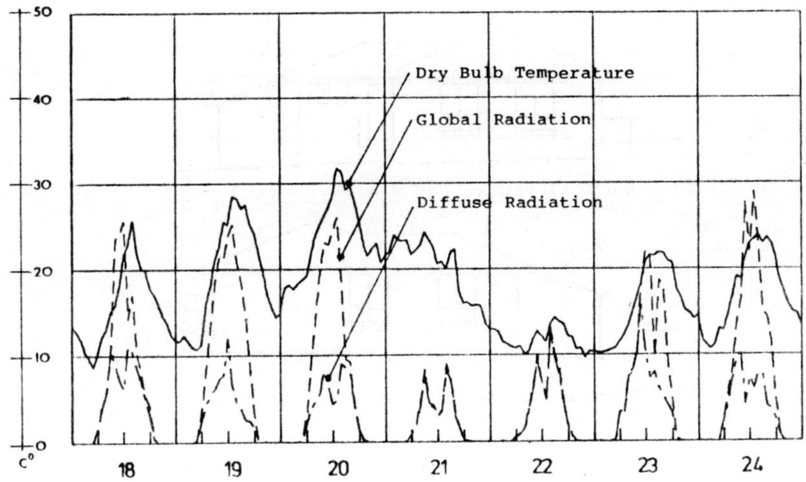

Fig. Al.2 External Climate, Broadmeadows

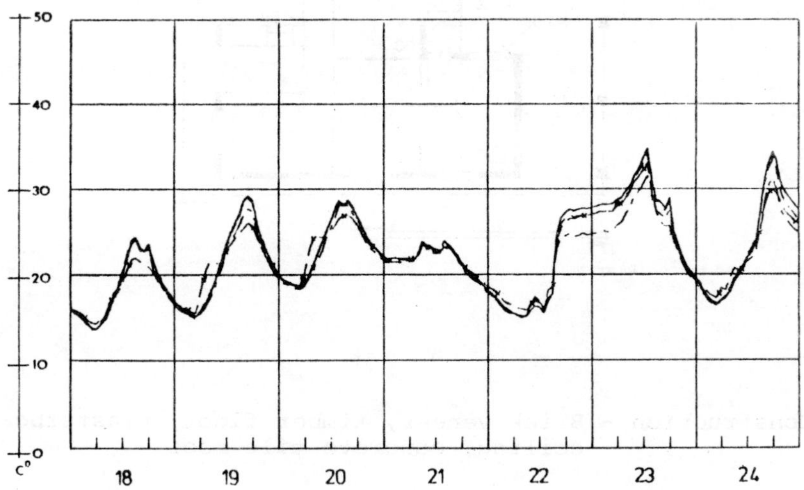

Fig. Al.3 Internal Environmental Temperatures, Broadmeadows

APPENDIX A (cont.)

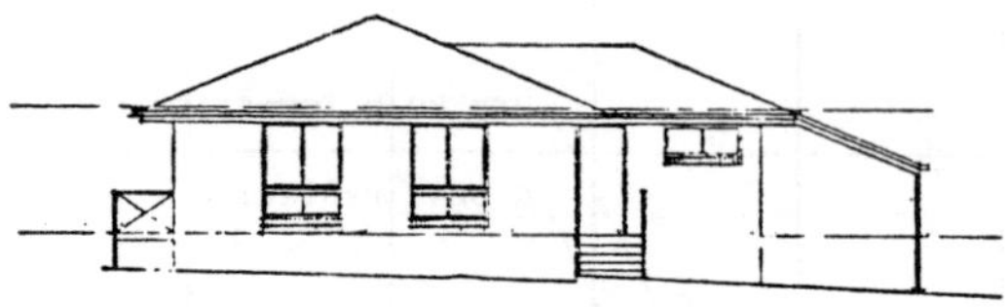

North elevation

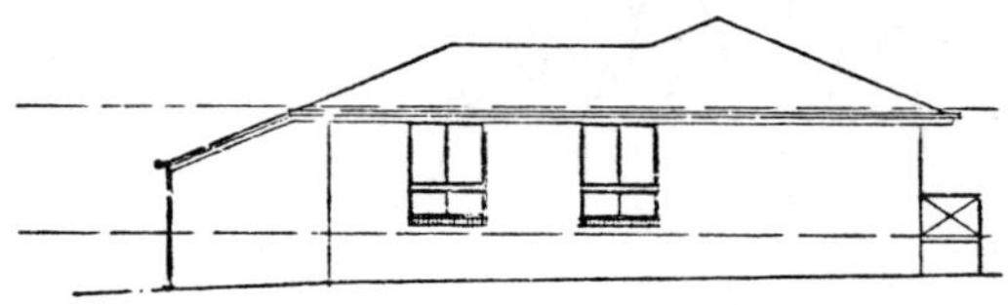

South elevation

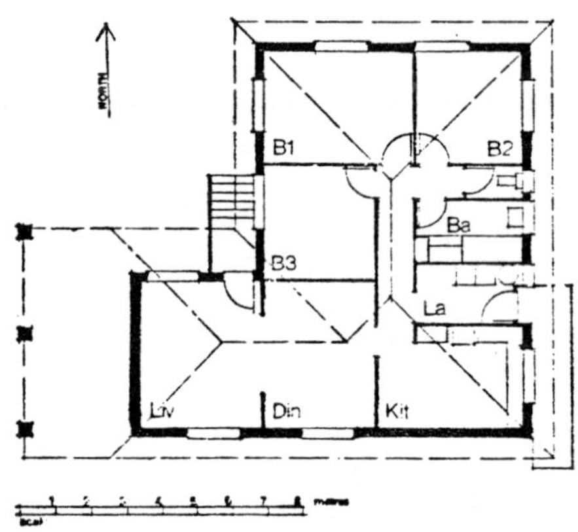

Plan

Construction - Brick veneer, timber floor, plasterboard ceiling, concrete tile roof

Fig. A2.1 House Monitored at Goodna, Brisbane

APPENDIX A (cont.)

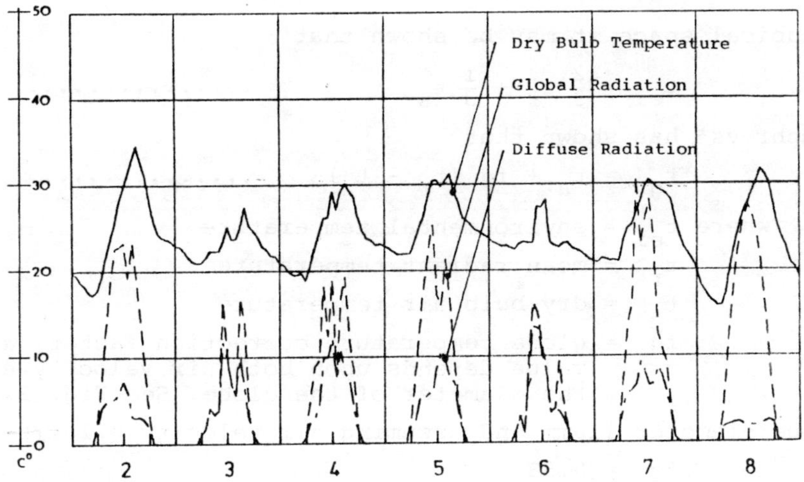

Fig. A2.2 External Climate, Goodna

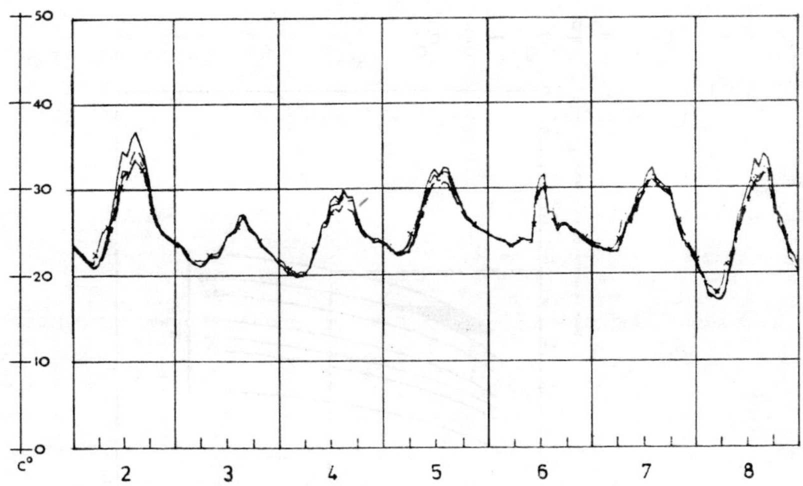

Fig. A2.3 Internal Environmental Temperatures, Goodna

APPENDIX B

Derivation of Relationship Between Globe Thermometer Temperature and Air Dry Bulb Temperature

For a cubical space it may be shown that

$$t_{ei} = \frac{2}{3} t_r + \frac{1}{3} t_a \qquad \text{...............} \qquad (1)$$

and Humphreys* has shown that

$$t_g = t_a + fg\,(t_r - t_a) \qquad \text{...............} \qquad (2)$$

where t_{ei} = environmental temperature

t_r = mean radiant temperature

t_a = dry bulb air temperature

fg = globe temperature correction factor; and value depends upon both air velocity and the diameter of the globe. See Fig. B1

For globe diameter 110mm and assuming air velocity 0.1 m/s

$$fg = 0.58$$

From eq.(2)

$$t_g = 0.58\, t_r + 0.42\, t_a \qquad \text{...............} \qquad (3)$$

Combining eqs.(1) and (2)

$$t_{ei} = \frac{6}{5} t_g - \frac{1}{5} t_a$$

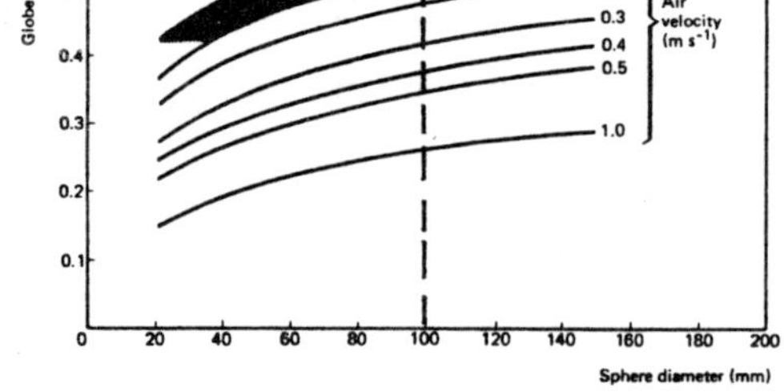

Fig. B1

---

* Humphreys, M.A. "The Optimum Diameter for a Globe Thermometer for Use Indoors," Am. Occup.Hyg., Vol.20, 1977, pp.135-140.

Some Aspects of Thermocirculation Heat Transfer in Trombe-Michel Walls

W.W.S. Charters
Reader in Thermodynamics, University of Melbourne
and Chairman of Victorian Solar Energy Council

INTRODUCTION

It is now becoming widely recognized that passive solar energy systems can contribute in a variety of ways to space heating and cooling requirements of relatively conventional building structures for domestic and commercial occupation. In these structures the thermal collection and storage is integrated into the building shell and thermal distribution is carried out by natural heat transfer processes thereby obviating the requirement for active distribution components such as fans or pumps.

There was an early type of solar space heating device (1) patented in the USA, as early as 1881 by Professor E.L. Morse of Salem (Massachusetts), which was essentially a free convection vertical flat plate collector with the air passage behind the plate as shown on Fig. 1. It is quite evident, however, that this device had only minimal storage of thermal energy and was possibly more useful in the ventilation mode.

In France, in 1956, a form of passive solar home heating system incorporating massive thermal storage was developed by Professor F. Trombe of the CNRS laboratory at Font Romeu-Odeillo in conjunction with a practicing architect Jean Michel. The resulting system has since become known as the Trombe-Michel Wall or often today this is shortened to the Trombe Wall. An early house, built utilising this principle at Font Romeu-Odeillo, used the complete South facing wall of the house as a solar collector thermal storage unit. Two aspects of this design provoked considerable discussion and adverse comment from architects: the first being the total dedication of the South face of the dwelling to solar collection, and the second the use of a matt black surface on the concrete for the glazed wall. Uncertainty was also expressed as to the desirability of using such massive thermal storage of the order of 0.67 m of concrete in this design.

The basic concept of the Trombe Wall has, however, now become widely accepted in many countries and has established itself as

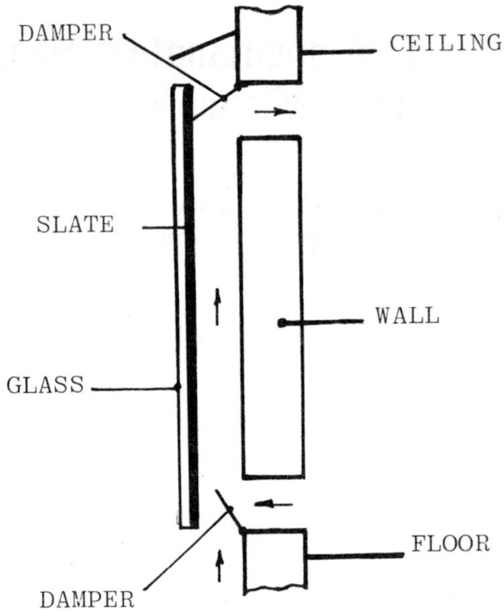

Fig. 1 Early Vertical Passive Solar Air Heater Morse (1)

Photo. 1 First Trombe-Michel House
Font-Romeu-Odeille, France

one of the many passive solar heating techniques now available to the practising engineer and architect.

OPERATIONAL PRINCIPLES OF A TROMBE-MICHEL WALL

The Trombe Wall uses a sun facing wall (South in the Northern Hemisphere or North in the Southern Hemisphere) which is normally vertical and forms an integral part of the building structure. This wall is then glazed in some form to allow the wall to act as a vertical solar collector by trapping solar energy using the well known greenhouse principle of an infrared opaque glazing material such as glass. The wall is normally constructed of brick or concrete although a wide variety of other materials has been proposed including adobe (mud brick). The main attribute of the construction material for this purpose is its ability to store thermal energy and to conduct the heat to the heated room space. The relevant thermophysical properties of some typical building materials are given in Table 1.

In addition to the conduction heat transfer through the storage wall air circulation vents are placed at the top and bottom of the wall thereby connecting the heated space with the Trombe Wall duct, Fig. 2. These vents allow the cooler room air to be drawn through the duct. The heated air is then delivered to the room through the top vents completing the thermal loop. This process of natural thermosiphon is often called thermocirculation in the literature and provides a direct path to the room without the extensive thermal delay inherent in the conduction path.

Ideally in practice these two heat transfer paths should be matched to provide comfort living conditions during a specified period of the day. If the wall is operated without use of the thermocirculation vents it is commonly called a thermal storage wall.

Because in practice the structural walls of a building are generally vertical the Trombe Wall is best suited to higher latitude locations where the winter sun is low in the sky. This allows the solar radiation to more readily penetrate the glazing avoiding the problem of reflection at low incidence angles and to fall more directly on the vertical wall. With the higher latitude locations the extensive change in the sun's trajectory during the year also allows wall shading by overhangs or eaves at seasons when heating is not required. As an example in Melbourne, Australia (latitude 38°S) the summer sun makes an angle of 76° to the horizontal plane at noon on '22 December at the summer solstice but only 28° to the horizontal at noon on '22 June at the winter solstice. This effect is illustrated for Melbourne on Fig. 2 and it can be seen that it is relatively easy to shade the wall entirely from summer sun.

TABLE 1 Thermophysical Properties of Some Typical Building Materials.

WALL MATERIAL	THERMOPHYSICAL PROPERTIES		
	Density (P) kgm^{-3}	Specific Heat (c_p) $kJkg^{-1}K^{-1}$	Thermal Conductivity (k) $Wm^{-1}K^{-1}$
Adobe (Mud Brick)	1730	0.84	0.70
Brick	1920	0.84	0.72
Concrete	2300	1.13	1.70
Water	1000	4.18	0.61

THERMAL PROPERTIES AFFECTING WALL PERFORMANCE	Thermal Capacity per Unit Material Storage Volume (ρc_p) $kJm^{-3}K^{-1}$	Thermal Diffusivity $(k/\rho c_p)$ m^2s^{-1}
Abode	1453	4.82×10^{-7}
Brick	1613	4.46×10^{-7}
Concrete	2599	6.54×10^{-7}
Water	4180	1.46×10^{-7}

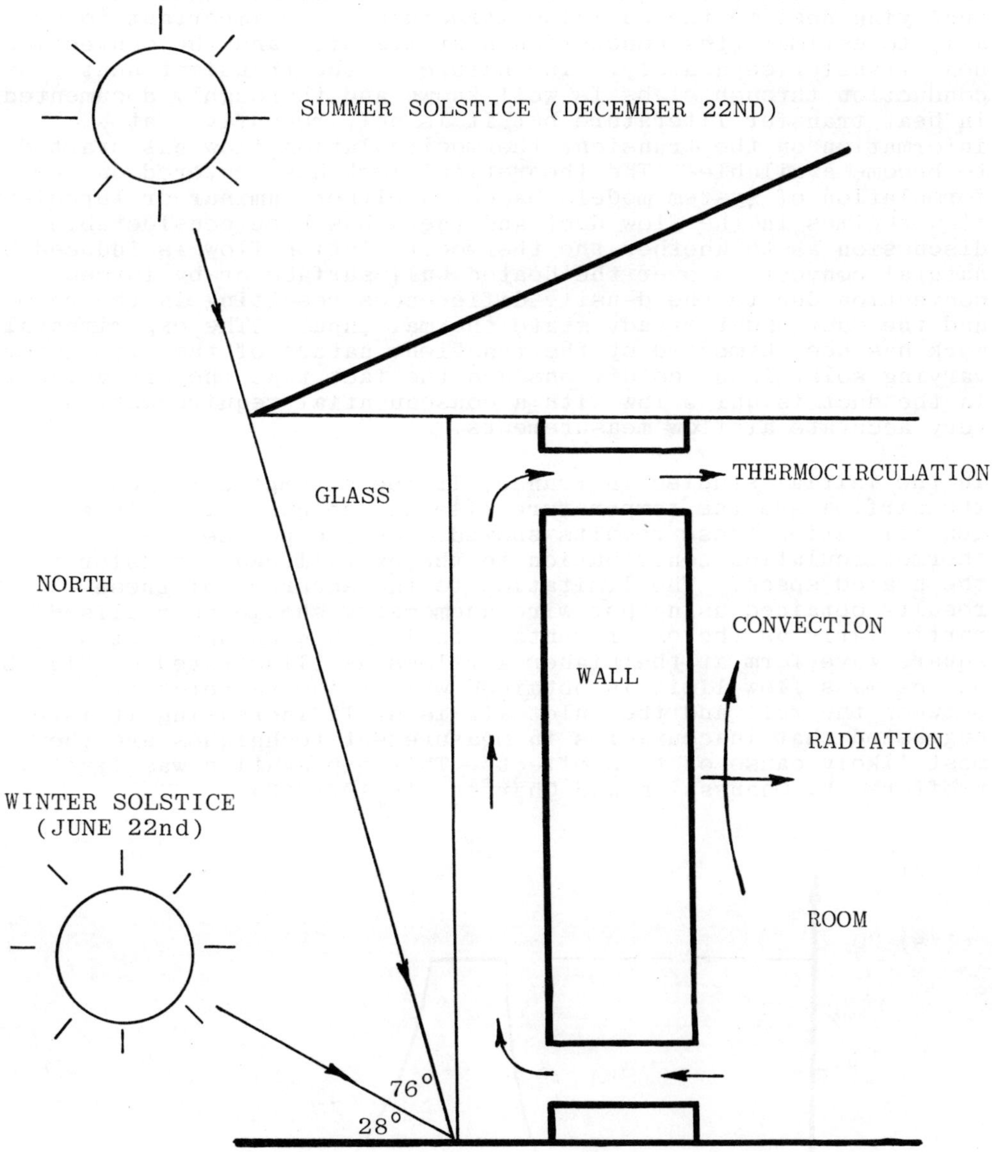

Fig. 2 Cross Section of Trombe Michel Wall Structure

THERMOCIRCULATION HEAT TRANSFER: THEORETICAL AND EXPERIMENTAL CONSIDERATIONS

In order to evaluate the usefulness of the Trombe Wall in supplying heat to the building structure it is important to be able to estimate the conduction heat transfer and the convection heat transfer separately. The nature of the transient heat conduction through slabs is well known and thoroughly documented in heat transfer literature but it is only recently that information on the transient thermocirculation flow has started to become available. The theoretical work has centered on the formulation of system models based on either laminar or turbulent flow regimes in the flow duct and there has been considerable discussion as to whether the thermocirculation flow is induced by natural convection over the heated wall surface or by forced convection due to the density differences resulting in the room and the duct under steady state thermal input. The experimental work has been hampered by the transient nature of the flow under varying solar input conditions and the fact that the air velocity in the duct is quite low with a consequential requirement for very accurate airflow measurements.

In the initial studies in France, Trombe attempted to measure the airflow and the temperature rise up the duct (2). As a generalisation these results showed about 30 per cent thermocirculation contribution to the overall heat transfer to the heated space. The limitation to the accuracy of these results obtained using hot wire anemometry should be realised particularly as the experimental results show an unexpected square wave form at the higher airflows as illustrated on Fig. 3. As the mass flow limit is obtained whilst the temperature between the wall and the inlet air is still increasing it is suggested that inaccuracies in measurement techniques are the most likely cause of this effect. This explanation was first proffered by Ohanessian and Charters in 1978 (3).

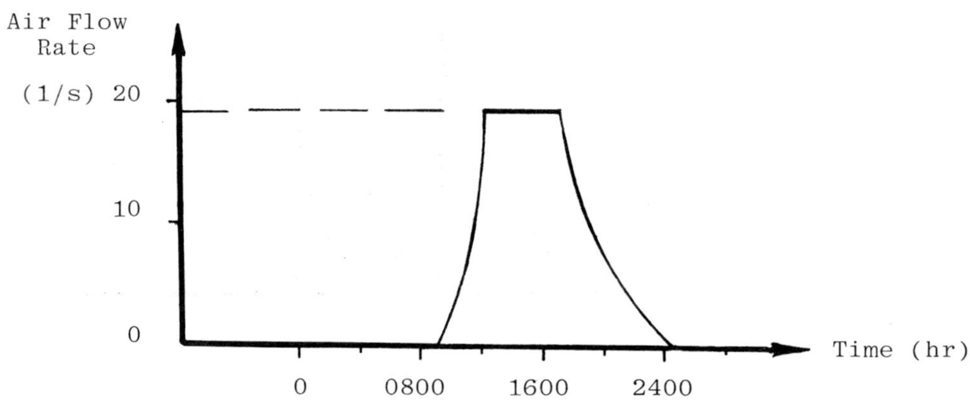

Fig. 3 Variations in Daily Airflow Rates Trombe et al (2)

Much of the work in the USA has been based on the simulation model developed by Balcomb et al (4) using a one dimensional implicit thermal network analysis to calculate room temperatures. This theoretical model, which has been validated using a series of simple test cell structures at Los Alamos Laboratories, New Mexico, assumes that the convective heat transfer in the wall duct can be treated as forced convection. The flow velocities are determined using a driving force derived from considerations of the differential weight of the air column in the room and the losses in the flow duct and the interconnecting vents to calculate the resulting flow velocities. A simplifying assumption made is that the losses in the duct are negligible compared with the flow losses in the vents resulting in the following equation.

$$\text{Air flow, } \dot{V}(m^3/s) = C_d\, A_v\, \sqrt{2\, gh\, B\, (T_O - T_I)}$$

for T_O, T_I — Outlet and inlet vent temperatures (K)

h — Wall length (m)

B — Volumetric coefficient of expansion of air (K^{-1})

g — Gravitational constant (m^2/s)

C_d — Discharge coefficient for two rows of vents in series with similar cross-sectional areas A_v (m^2).

For this work the value of C_d has been calculated using a sudden contraction loss of half velocity head and a sudden expansion loss of 1 velocity head followed by a recovery from the vena contracta prior to the subsequent expansion (4). Using this technique the value of C_d becomes 0.57 for the wall structure.

As explained earlier the heat transfer is assumed to be that for forced air flow over the wall and glazing surfaces and the commonly accepted expression for the heat transfer coefficient in the form

$$h_c = 5.68 + 4.1\, V\ (Wm^{-2}\, K^{-1}) \qquad V\text{-uniform duct velocity } (ms^{-1})$$

was used throughout this modelling work, based on the assumption that the heat transfer coefficient can be taken constant over the small range of flow velocities encountered in the flow duct. By then using a thermal balance equation for the flow in the duct it was possible to theoretically determine the circulation heat transfer from the following equation

$$\dot{Q} = \rho \dot{V} c_p\, (T_O - T_I) = h_c\, A_W\, \{(T_W + T_G) - (T_O + T_I)\}$$

for c_p — Specific heat of air ($Jkg^{-1}\, K^{-1}$)

A_W Wall surface area (m^2)

T_G, T_W Glazing and wall temperatures (K)

When the prediction results were compared with experimental results from the test cells the temperature predictions for the heated space showed a maximum error of 5 per cent at the times when the thermocirculation was most intense. This is claimed to be an acceptable degree of accuracy as the total amount of heat transferred to the room by convection is less than half of that transferred by the transient conduction mechanism through the wall.

Other theoretical studies which throw light on the possible nature of the thermocirculation flow include those by Akbari and Borgers (5), (6) in which analyses of the duct flow have been carried out for both laminar and turbulent flow regimes, assuming natural convection flow between parallel vertical plates. For these studies the transition criteria have been based on the airflow rate and the duct width, and the vent loss coefficients have been assumed to be unity. The full equations resulting from these studies are both lengthy and complex for general use but the investigations do aid in the understanding of the complex flow phenomena in these deceptively simple wall structures. Another study by Pratt (7) involves the use of a slightly different numerical technique in that the duct model proposed is idealised to consist of an entrance/orifice, a flow duct section, followed by a second orifice/exit. Once again the flow is considered to be of a two dimensional nature in between vertical flat plates held at unequal but uniform or isothermal surface conditions. In order to evaluate the head loss coefficients appertaining to the assumed square edged orifices previous experimental work on duct flow is used. Uniform velocity profiles are assumed at inlet and exit to the identical duct, Fig. 4.

A limitation on this numerical model is that the head loss coefficients should be limited to values less than 2.0 to ensure that there is no downward airflow along the glazing surface. In this theoretical study calculations were presented for the wall and the glazing heat transfer coefficients and it is interesting to note that over a relatively wide range of duct width and glazing temperatures the heat transfer from the wall surface to the air stays essentially constant. It is, however, noted in this work that it is possible to increase the heat transfer from the wall to the floating air stream by using less restrictive vents.

Some recent experimental work in the USA by Casperson and Hocevar (8) was carried out on a test cell incorporating full width wall vents. The wall was double glazed to take account of the lower ambient temperatures encountered in North American conditions. Interesting results which can be interpreted from the duct velocity and temperature traverses indicate the

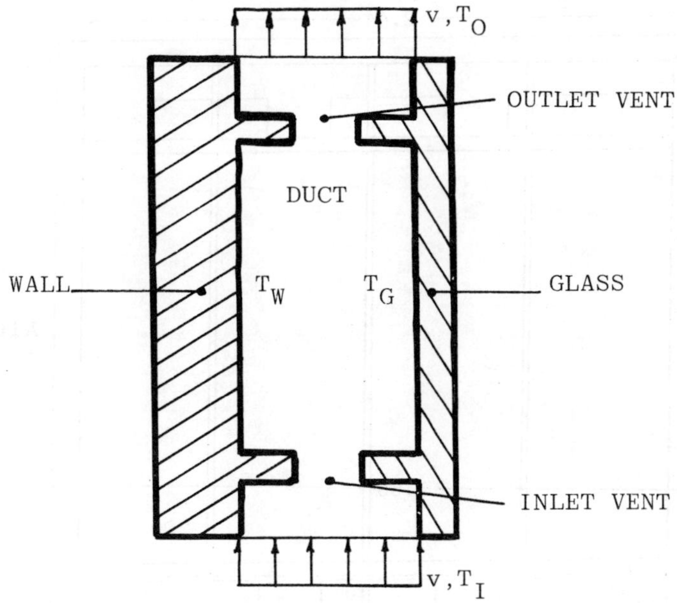

Fig. 4 Idealised Duct Configuration Pratt (7)

existence of two boundary layers. The attachment of a boundary layer to each of the vertical plate surfaces is indicative of natural convection flow in the duct. Although the authors stated that the flow was felt to be laminar no experimental validation of this statement was produced, although the general trend of the velocity profiles was in line with the theoretical laminar predictions of Akbari and Borgers op.cit. Some attempt was made to differentiate the thermal performance of the wall for duct widths varying from 50 mm to 150 mm but no outstanding performance changes were noted. As the large thermal capacity of the Trombe Wall virtually ensures that the thermal behaviour is continually transient any assumption of steady state methods for analysis should be treated with caution. Other experimental results reported recently include those by Akbarzadeh et al (9) from investigations at the University of Melbourne, Australia. The test cell in this case was fitted with a single glazed assembly capable of movement to alter the wall/glazing distance from 100 mm to 350 mm to allow investigation of the heat transfer and airflow characteristics with differing duct configurations, Fig. 5. All of the airflow measurements were taken using a wind tunnel calibrated hot film anemometer because the low air velocities, the dusty environment, and the essentially unsteady nature of the flow precluded the use of other measuring devices such as laser doppler velocity meters. After initial flow visualisation studies using heated metaldehyde (a solid camping fuel) shaped vent inserts were incorporated into the top vents resulting in a reduced flow unsteadiness and a uniform flow velocity across the plane of measurement. All the adjustments required for investigating the effect of vent size on thermocirculation were made on the

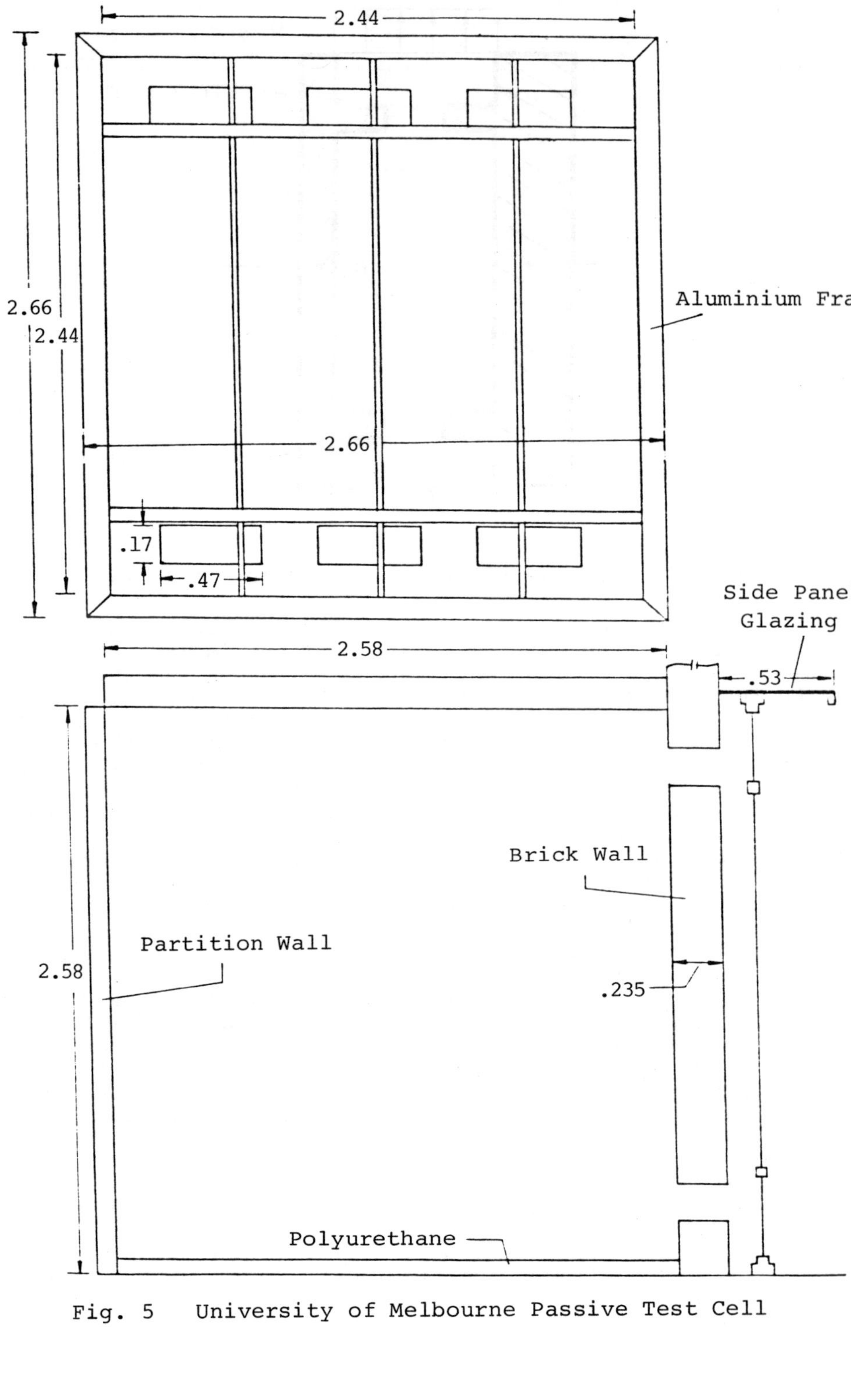

Fig. 5 University of Melbourne Passive Test Cell

bottom vents only. The claimed accuracy of the airflow velocity measurements using the hot film anemometer are ± 10 per cent.

TYPICAL FLOW PATTERNS IN TROMBE WALLS:

Although the following comments are based on observations made from a particular installation, namely the passive test cell at the University of Melbourne, they appear to be of a general enough nature to be widely applicable to flow situations of this type.

As the air enters the duct from the inlet vents some of the air is turned downwards and some upwards after issuing from the vents and eddies are formed along the sides and beneath the vents. For many test runs a separated flow region of about 100 mm length was observed just above the inlet vent. For the top vent connecting the wall back to the heated space the air entered the vent both from above and below the vent orifice, Fig. 6, 7.

Along the entire length of the duct the flow was observed to be turbulent with the degree of turbulence higher in the vicinity of the vent openings suggesting an augmentation of the turbulence due to the sudden expansion and contraction of the airflow in these regions. The highest upward velocities were always observed closest to the highest wall temperature region next to the blackened brick surface with the bulk of the duct flow passing close to this surface, Fig. 8.

The airflow patterns on the glazed surface are more complex than those over the heated wall surface. For example during the day the flow direction can vary substantially. During early morning and late afternoon tests it has been noticed that downward airflows exist over the entire glass surface. It would appear that at these times when the inlet air temperature exceeds the glass temperature that the glazing has a cooling effect on the air resulting in this reverse flow situation. During the day as the ambient air temperature rises with increasing solar radiation the glass temperature also rises until it exceeds the temperature of the duct inlet air. At this point the glazing can start to heat the inlet air and an observable upward motion of air over the lower glazing results, although there may still be a downward cell of air at the top glazing if the bulk temperature of the upper airstream exceeds that of the glass. This downwash or reverse flow effect will always be in evidence if the temperatures in the top glazed region do not attain sufficiently high values to counteract it.

Comments on the Mechanism of Heat Transfer in the Duct:

From some of the theoretical and experimental evidence recently appearing in the literature it seems reasonable to conclude that the heat transfer processes occuring within the duct of a Trombe Wall can be adequately modelled using turbulent natural convection between independent vertical isothermal flat plates. This 'dual plate' model is supported by visual observations of the flow phenomena as well as experimental results. As with

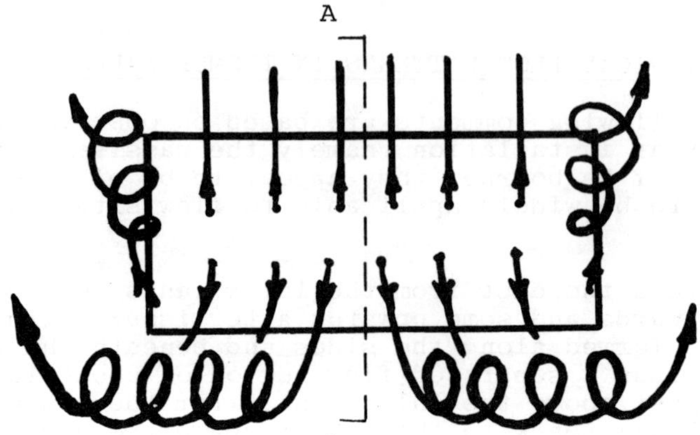

Fig. 6 Simplified Airflow Patterns - Inlet Vent

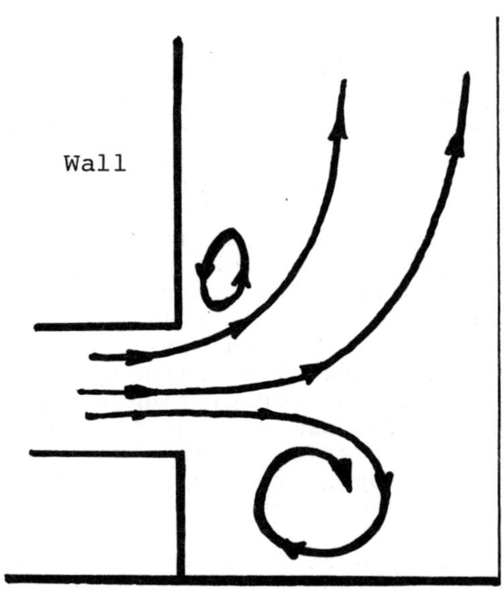

Section AA

Fig. 7 Simplified Airflow Patterns - Inlet Vent

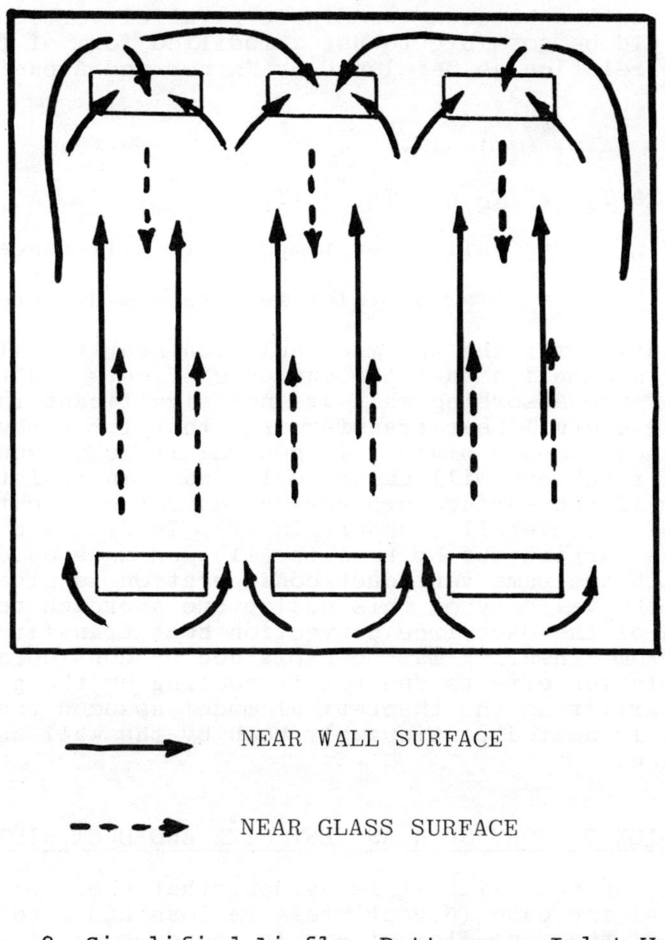

Fig. 8 Simplified Airflow Patterns - Inlet Vent

all models differences will exist between reality and the idealised concept. For example, the flow will tend to have a three dimensional rather than two dimensional nature as the vents do not in practice extend the full width of the wall module. Because of the restrictive nature of the vents there will be a higher overall pressure drop and a lower mass flow of air than for a flat plate in an infinite medium. In addition the wall surface in general will not be isothermal as the top wall region is at a higher temperature than the lower region because of heat transfer effects. If as seems probable from recent studies that the flow in the Trombe Wall duct can be taken as turbulent natural convection between two single parallel vertical plates then it should be possible to use a modified form of the heat transfer correlation as developed by Warner and Arpaci (10) in the form

$$Q_t \propto (\Delta T_{WI}{}^{4/3} + \Delta T_{GI}{}^{4/3})$$

where $T_W > T_G$ and $T_G > T_I$

ΔT_{WI} = Wall inlet temperature difference

ΔT_{GI} = Glazing inlet temperature difference.

This formulation for the thermocirculation heat transfer is based on an assumption that the entrance effects and the surface roughness of the absorbing wall are not significant factors in altering the overall heat transfer from that for turbulent flow over a vertical smooth plate. In general it is reasonable to assume that roughness will cause no increase in resistance to the airflow if the surface protrusions do not pierce the laminar sub layer of the overall boundary layer. In Fig. 9 a comparison of many runs carried out by Lesslie (11) and made on different days but with the same vent/duct configuration is given as an example of the validity of this dual plate approach to the calculation of the duct free convection heat transfer. Some deviation from linearity may be explained by consideration of the recirculation effects due to air cooling by the glazing discussed earlier as the theoretical model assumed that the air in the duct is heated continuously both by the wall and the glass surface.

DESIGN FACTORS SUCH AS VENT SIZE AND DUCT WIDTH:

By reducing the vent size it is evident that there will be a consequential increase in vent pressure loss and a corresponding diminution of the mass flow of air in the duct. This reduced airflow will result in an increased temperature rise up the duct, which would be expected to increase the possibility of recirculation down the front glazing. This effect has been pointed out by several workers who have stated that recirculation appears to be inevitable for very restrictive vents. From the work at the University of Melbourne it appears that a substantial reduction of vent size from 170 mm depth to

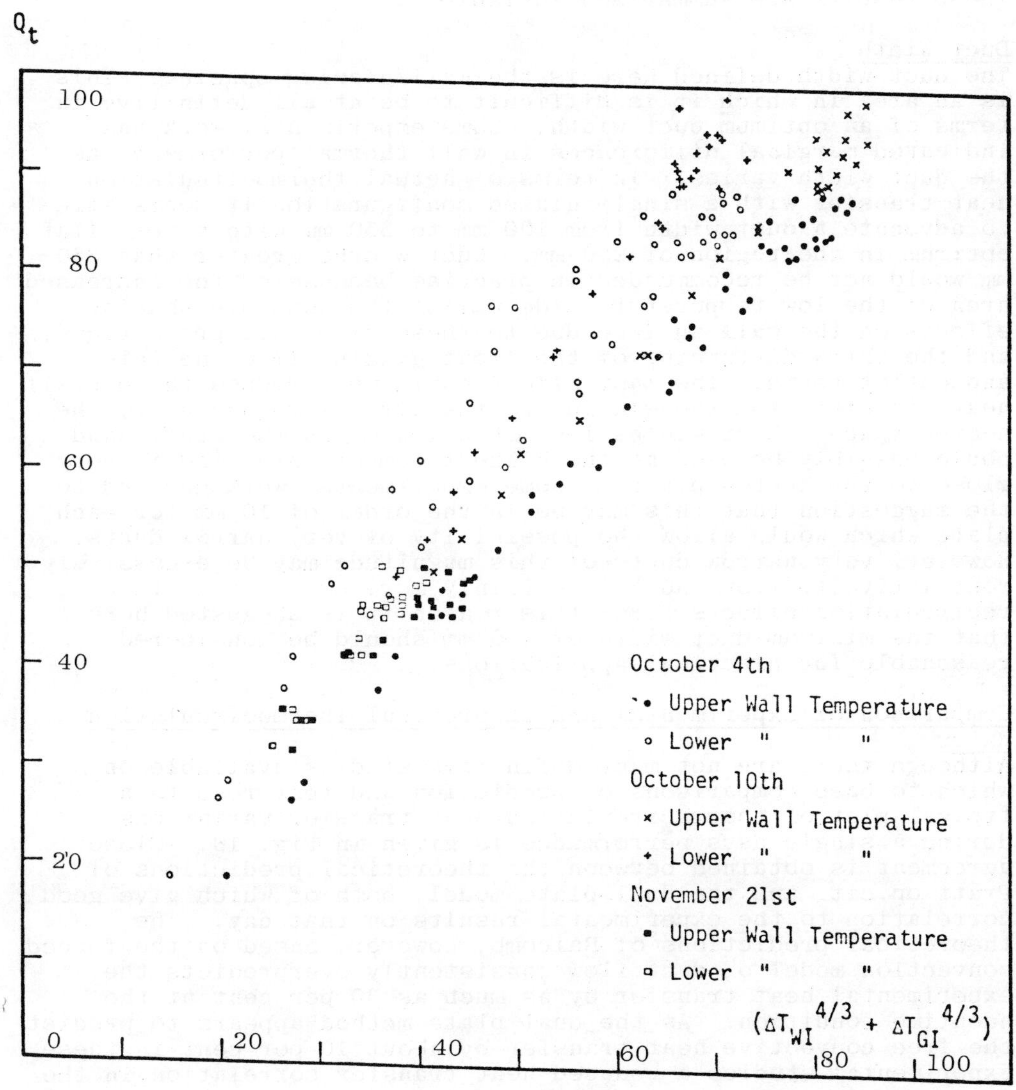

Fig. 9 Experimental Test Results for Different Climatic Conditions Lesslie (11)

45 mm was sufficient to cause a major reduction in the convective heat transfer but that a reduction to 95 mm provided no significant change in this parameter. In the latter case it is thought that the reduction in airflow was equally offset by the increased duct temperature rise balancing out the effect. These results are summarised in Table 2.

Duct Width

The duct width defined here is the wall/glazing spacing. This is an area in which it is difficult to be at all definitive in terms of an optimum duct width. Some experimental work has indicated marginal differences in wall thermal performance as the duct width varies. In terms of actual thermocirculation heat transfer with a single glazed configuration it seems safe to advocate a duct width from 100 mm to 350 mm with a very flat optimum in the region of 250 mm. Duct widths greater than 350 mm would not be recommended in practice because of the increased area of the low temperature side walls, the possible shading effects on the wall surface due to these side wall projections, and the extra distancing of the front glazing from the inlet and outlet vents. The last effect could well reduce the overall heat transfer from the glazing to the airflow returning to the heated space. Duct widths less than 100 mm on the other hand could possibly be used as the highest temperature rise is very close to the heated plates: some experimental work has led to the suggestion that this may be in the order of 10 mm for each plate which would allow the possibility of very narrow ducts. However, very narrow ducts of this magnitude may be excessively restrictive to flow and are certainly more susceptible to recirculation effects. For this reason it is suggested here that the minimum duct width of 100 mm should be considered reasonable for practical applications.

Comparison of Experimental and Theoretical Thermocirculation Rates

Although there are not many definitive studies available on which to base comparisons of prediction and test results a typical plot of thermocirculation heat transfer variations during a single days performance is given in Fig. 10. Close agreement is obtained between the theoretical predictions of Pratt op.cit. and the dual plate model, both of which give good correlation to the experimental results on that day. The theoretical predictions of Balcomb, however, based on the forced convection model of duct flow consistently overpredicts the experimental heat transfer by as much as 30 per cent at the noontime condition. As the dual plate method appears to predict the free convective heat transfer by about 10 per cent in these experimental studies a reduced heat transfer correlation in the form,

$$Nu = 0.09\ Ra_L^{1/3}$$

is suggested as adequate for general design purposes in most practical applications where the suggestions regarding vent size and duct width are followed.

TABLE 2 Effect of Vent Size on Thermocirculation Heat Transfer.

Test Results - University of Melbourne Passive Test Cell

October 4th (D = 35cm, V = 17cm)			October 10th (D = 35cm, V = 9.5cm)		
Q_t (W/m^2)	$\dot{m}$ (kg/s)	ΔT_{OI} (^oC)	Q_t (W/m^2)	$\dot{m}$ (kg/s)	ΔT_{OI} (^oC)
61.1	.072	5.97	59.9	.064	6.60
63.6	.070	6.38	61.6	.062	7.00
60.8	.069	6.22	61.5	.061	7.12
59.1	.067	6.19	60.9	.061	7.09
			64.5	.065	7.04

Mass flow rates and temperature rises up the duct for similar thermocirculation heat transfers for 4 October and 10 October, 1980.

October 4th (D = 35cm, V = 17cm)			October 22nd (D = 35cm, V = 4.5cm)		
Q_t (W/m^2)	$\dot{m}$ (kg/s)	ΔT_{OI} (^oC)	Q_t (W/m^2)	$\dot{m}$ (kg/s)	ΔT_{OI} (^oC)
67.4	.074	6.39	70.0	.047	10.52
72.5	.076	6.72	68.3	.047	10.29
69.5	.073	6.70	69.3	.047	10.43
68.8	.072	6.75	66.9	.047	10.12
			68.7	.047	10.41
			68.0	.046	10.46
			68.4	.046	10.42

Mass flow rates and temperature rises up the duct for similar thermocirculation heat transfers for 4 October and 22 October, 1980.

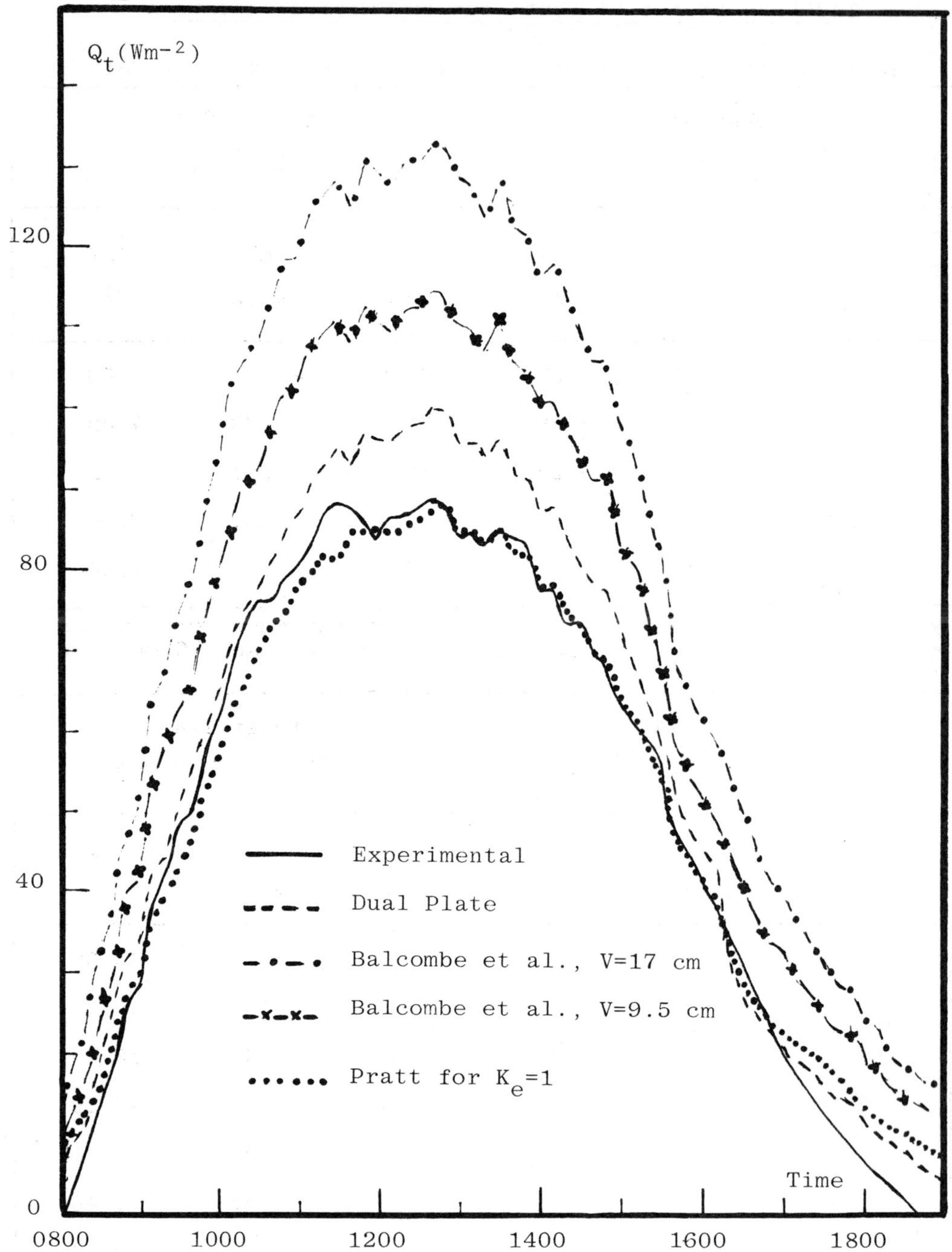

Fig. 10 Comparison of Theoretical and Experimental Results for Thermocirculation Heat Transfer Akbarzadeh et al (9)

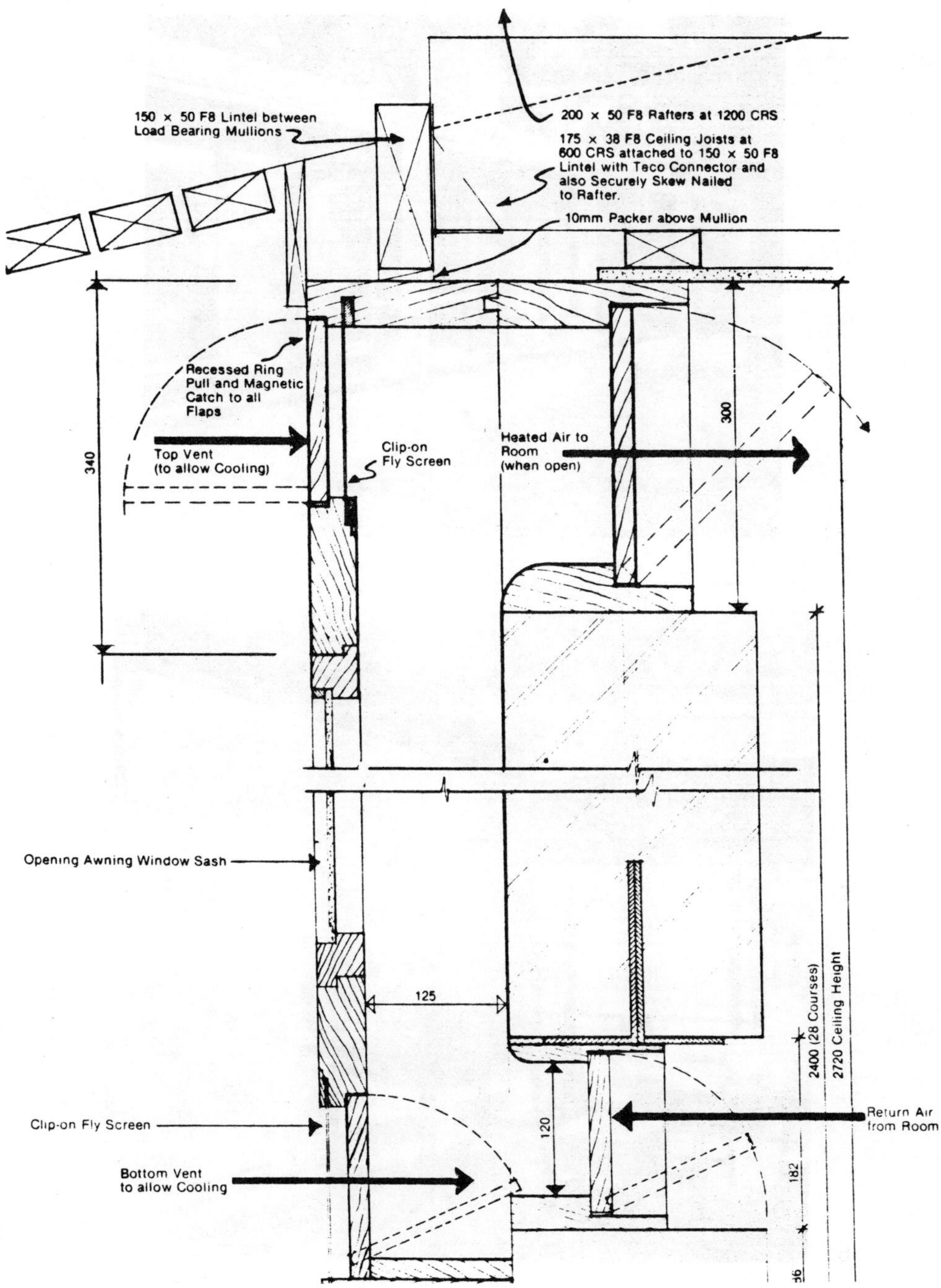

Fig. 11 Details of Vent Construction for Merchant Builders Trombe House Melbourne, Australia.

(a)

(b)

Photo 2 Merchant Builders Trombe House, Melbourne, Australia: (a) exterior, (b) interior

Practical Construction of Trombe Walls in Dwellings:
As a follow on to the work already described at the University of Melbourne a major project has been undertaken by the University in conjunction with Merchant Builders Pty. Ltd. This involves the building of two houses of identical floor plan and similar in all respects except for the addition in one house of Trombe Wall elements on the North face of the building. In the standard house this wall is of normal brick veneer construction, but in the Trombe house the glazed wall elements are of double brick construction. These houses will be monitored for a one year period, starting in June 1982, to assess the effectiveness of the Trombe Wall elements in reducing the overall energy consumption in the house. Some details of the vent duct configuration can be identified from the sketch on Fig. 11. It is hoped that the extensive monitoring of thermal flows in these test houses will lead to a fuller understanding of the complex and interactive heat transfer associated with this deceptively simple type of passive solar collection device.

ACKNOWLEDGEMENTS

The basic research work at the University of Melbourne on free convection effects was funded by the Australian Research Grants Committee and the development of the Trombe Wall modules by the Victorian Solar Energy Council and the National Research, Development and Demonstration Council.

REFERENCES

(1) E. L. Morse, US Patent No. 246626 1881, Warming and Ventilating by the Suns Rays.

(2) F. Trombe, J. F. Robert, M. Cabanat, and B. Sesolis, Some Performance Characteristics of the CNRS Solar House Collectors, Passive Solar Heating and Cooling Conf. Workshop. and Proc., Alberquerque New Mexico, 201-202, May 18-19, (1976).

(3) P. Ohanessian, and W. W. S. Charters, Thermal Simulation of a Passive Solar House Using a Trombe-Michel Wall Structure, Solar Energy, 20, 275-281, (1978).

(4) J. D. Balcomb, J. C. Hedstrom, and R. D. Macfarland, Passive Solar Heating of Buidlings, Los Alamos Scientific Laboratory Report. LA-UR-77-1162, June (1977).

(5) H. Akbari, and T. R. Borgers, Free Convective Laminar Flow Within the Trombe Wall Channel, Solar Energy, 22, 165-174, (1979).

(6) H. Akbari, and T. R. Borgers, Free Convective Turbulent Flow Within the Trombe Wall Channel, - unpublished, personal communication.

(7) R. G. Pratt, Turbulent Free Convection Flows in Trombe Walls, M.S. Thesis, Colorado State University, (1980).

(8) R. L. Casperson, and C. J. Hocevar, Experimental Investigation of Trombe Wall Passive Solar System, Proc. 3rd National Passive Solar Conference, San Jose, California, 231-237, January, (1979).

(9) A. Akbarzadeh, et al, Thermocirculation Charateristics of a Trombe Wall Passive Test Cell, accepted for publication, Solar Energy, (1982).

(10) C. Y. Warner, and V. S. Arpaci, An Experimental Investigation of Turbulent Natural Convection in Air at Low Pressure Along a Vertical Heated Flat Plate, Int. Jnl. Heat and Mass Transfer, 11, 397-406, (1968).

(11) D. A. Lesslie, Aspects of Trombe Wall Thermocirculation Heat Transfer, M.Eng.Sci. Thesis, University of Melbourne, Australia, (1981).

Computer Simulation of the Thermal Performance of Western Australian Housing Compared with Field Measurement

D.D. Carruthers
School of Architecture,
University of Western Australia

INTRODUCTION

A reliable computer program for thermal performance prediction is naturally of great value in the design of energy efficient buildings. Particular interest has been shown in recent years in the prediction of the performance of unoccupied, unconditioned buildings, especially with current interest in 'passive solar' architecture.

The program, to be useful, must behave reliably, be easy to operate, be capable of re-configuration to allow for changes in the building parameters at the design stage, and preferably not be too expensive to run. It is usually assumed that climatic data will be available in enough detail to enable sensible prediction to be made, an assumption seldom justified in practice, and that parameters whose values are not known with any degree of certainty will not be sensitive elements in the simulation. It is shown that this latter assumption, in particular, is in doubt.

COMPUTER PROGRAMS

A program to be of sufficient accuracy for passive simulations must handle transient phenomena well. There exists a large body of theoretical analysis which, in principle, enables such calculations to be undertaken, but many require considerable computer time.

The response factor methods developed under ASHRAE by Mitalas, Stephenson and others, have latterly been modified so that the behaviour of a building element can be categorised by a set of 'b' and 'd' coefficients of a series representing, in even steps, the hour by hour behaviour of that element. The elements are computed by a double Z-transform, and while not considering the long-term thermal history of the building, should be reasonably accurate.

One of the programs tested in this study was of this type. The program TRNSYS was developed by the University of Wisconsin, primarily for the analysis of active solar systems. It initially had a very crude model of the building itself, but later versions use the ASHRAE techniques mentioned above, albeit orientated to North American constructional practices. The version used here was the latest available when the project was initiated in 1979, namely Version 9.1.

It was quickly found that several features were insufficient for the current simulations. In particular, the following:

(a) Roof Simulation. Version 9 had incorporated into it parameters for a variety of roof/ceiling types not common in Australian practice. In consequence, a new subroutine was incorporated which enabled 'b' and 'd' coefficients computed using the program ZFORMD, made available by Mitalas and Arsenault (Ref. 1) at the Canadian Building Research Division, to be input as data.

(b) Window Shading. The incorporation of window shading is difficult in programs of this type. It can be in terms of a sunbreak ratio, but this rarely allows for windows of differing heights on one face, or, as here, merely a 'percentage shading'. In the project undertaken here, this was felt to be unsatisfactory, and in the data acquisition the actual percentage shading on each window was measured and input as data on an hourly basis. Data for a particular house was collected for 10-14 days, so that on a day near the middle of the period, a series of hourly photographs were taken.

(c) Window Transmission. The Transmission properties of the window glass were modified using another subroutine to allow for the lower transmission at high angles according to the Lorentz-Lorenz relation.

(d) Ventilation. The usual input of data is a constant assumed air change rate in each zone throughout the test period. In the modified TRNSYS, hourly wind speed was input as data, and the ventilation computed according to an empirically determined linear relationship between ventilation and wind speed. No account was taken of wind direction, although this was measured, and could be important especially as most houses had predominantly N and S windows, rather than an even distribution.

In most simulations, windows were assumed closed as this was the experimental procedure. In some tests, windows were opened, usually at night. A subroutine allows the actual ventilation in a.c.h. to be input as data and used to override the usual computation. The choice of ventilation value is then somewhat eclectic, and was normally taken as a constant 25 a.c.h.

Roof space ventilation is an even more uncertain quantity, but luckily research such as that by Van Straaten implies that the degree of ventilation of the roof space is not a major influence on ceiling temperature in comparison to roof temperature, and hence radiation transfer, at least when the major problem is one of overheating, as is the case in the localities illustrated in this paper.

The second program tested was ZSTEP, developed by the Division of Building Research, CSIRO, Highett, Victoria. This is an altogether more sophisticated program in its treatment of heat flows. The response is for the whole building and no part of the building can be altered during the run. It will deal readily with buildings in up to ten zones, and has better window shading routines. It has several drawbacks: The roof element must be represented by the equivalent flat roof, whereas TRNSYS allows for pitched roofs; ventilation must be maintained constant throughout the simulation period, since there are harmonic components in the simulation of several days duration. As a research program, it was difficult to input the data, but this was overcome at the University of Western Australia by the writing of an input program ZINPUT, by Roy et al (Ref. 2). The data required was similar to that for TRNSYS, except that cloud cover was incorporated into the sol-air temperature model, which latter was much more complex than the simple version in TRNSYS.

Data input was in a modified form of the Australian Bureau of Meteorology format. In our simulations, cloud cover in okta had not been obtained, so that inspired guesses based on the solar data as a guide were used. An attempt was made to relate the cloud cover to the known values of global and diffuse radiation or their ratio from Bureau of Meteorology data where okta values were given, but no useful correlation was found.

INPUT DATA

TRNSYS required the following climatic data:

(a) Global and diffuse solar radiation.
(b) Wind speed.
(c) External dry bulb temperature.

In addition, for comparison purposes, the temperature at the centr of each room was input. As the house was treated as one zone, the temperature in each room was multiplied by the floor area, and a mean zone temperature obtained by dividing the total of these figures by the total internal floor area. Experimental results showed that except where direct solar penetration occurred, the difference in temperature between rooms was small, no greater than the layering effect within one room, and usually less than 1°C.

As indicated earlier, the percentage shading of the windows on each face was also input. The parameters to be input include areas of each fabric on each face, together with their 'b' and 'd' coefficients for up to 6 hours, orientation, date, the absorptivity for short wave radiation and emissivity for long wave of all surfaces, the empirical equation for ventilation, and ground reflectance. There is always some uncertainty on the values to be used for solar absorptivity and ground reflectance, which former can be a problem, especially with an uninsulated roof.

ZSTEP requires the extra climatic data of cloud cover as indicated above, and fabrics are specified in terms of their thermal resistance and capacitance as the transient coefficients are calculated within the program. Full details of the input procedure for ZSTEP are given in Reference 2.

EXPERIMENTAL DATA

No attempt will be made in this Paper to elaborate on the methods used to obtain the data except the following points:

(a) Wind speed was measured at the most remote point on the house from the house itself, at eaves height. There is a conflict in that the wind speed is required within the programs, both to estimate external air film resistance on the roof and to compute ventilation. The building itself will grossly modify the wind speed both on the roof and at the windows, and no attempt was made to obtain a wind profile down a vertical wall as this would be a continually varying quantity and could not reasonably be incorporated into the ventilation computation.

(b) All temperatures were measured using small copper constantan thermocouples, with cold junction compensation provided by the logging equipment. The compensation was checked using pure melting ice before and after each run, but not during the run.

Other data not used as input included heat flows through surfaces, surface temperatures and wet bulb temperatures. The empirical ventilation equation for each house was obtained using the usual nitrous oxide injection and an infra red gas analyser. Where ceiling fans were fitted, they were run before and during the tests.

Test Houses

Of the seven test sites measured during this project, results will be discussed for two, both warm to hot climates.

South Hedland. This site has hot weather in December-March with a pleasant warm winter. The houses here were of veneer construction, with external concrete block leaf and an inner stud wall of 'Gyprock'. The wall cavity carried reflective foil insulation on the outside of the stud. The inner walls were studwork and 'Gyprock', the ceiling 'Gyprock' and the roof metal deck of a buff colour. The floor was concrete slab on ground. The surrounding area was bare earth of a red-brown colour.

Geraldton. This site has an equitable warm climate, modified by close proximity to the sea. The houses here were, in fact, one half of a duplex ('semi-detached') house, of brick veneer construction, stud inner walls, 'Gyprock' ceiling and a cement-tile roof of red colour. The houses had internal floor area of about 100 m^2.

RESULTS

In analysing the results, one clearly wishes to ascertain to what extent the temperature regime of the experimental house is followed by the simulation. For clarity, the performance can be discussed in terms of mean temperature and the lag and decrement of the mean indoor dry bulb temperature. Table 1 shows the day by day analysis of two houses in Geraldton which are identical except that House A has an insulated ceiling. The mean temperatures for the 12 day period, not shown in the Table, are:

House A:		House B:	
EXPERIMENT	24.9	EXPERIMENT	24.4
TRNSYS	22.6	TRNSYS	23.4
ZSTEP	23.7	ZSTEP	24.3

Several features should be immediately noted. First, it must be remembered that the temperatures in ZSTEP results are not strictly dry bulb temperature, but a combination of mean radiant temperature and air temperature. In the context of these results, the difference is likely to be small, as air temperature, wall temperatures and globe temperature are very similar. A second feature is that throughout the tests, the uninsulated ceiling was simulated better than the insulated. This could be partly due to some uncertainty as to the values of thermal constants for the fibreglass used, but is more subtle than that, as measurement of heat flow through the ceiling showed good agreement between theory and experiment in the case of the TRNSYS simulation, in which the agreement in the insulated case is not good.

Table 1 and Fig. 3 show TRNSYS tends to predict minima well, but that ZSTEP gives better agreement on maxima. It must be remembered that the cloud cover at night was a subject of conjecture. If more cloud were present at night, the ZSTEP night figures would be higher. The thermal lag on maxima is better with ZSTEP, and the TRNSYS plots are too highly damped, so that decrements are always too low. ZSTEP decrements are too high, but by an almost constant factor. In the case of the 'B' house this factor is between 1.3 and 1.5.

A plot of 2 days data shows the comparison of programs very clearly. The excessive dip in ZSTEP is well seen and the effect of reducing the thermal mass assumed in TRNSYS, which raises the maxima while leaving the minima little affected, is also clear. ZSTEP is seen to give a better simulation of the daily variation in this case.

The <u>South Hedland</u> results are shown in Table 2 and Fig. 1 for winter and in Table 3 and Fig. 2 for summer. The decrement agreement is quite good for ZSTEP, and very much better for TRNSYS. TRNSYS handles the thermal lag much better than ZSTEP, especially in winter, but the reverse is true in summer. The comparison is again clearly seen in 2 day plots, where it can be seen that ZSTEP simulates well for Hedland 'A' in winter, except that the curve is offset downwards.

In the uninsulated house, ZSTEP is superior in all except lag, TRNSYS having the correct shape, but being offset to too low temperatures.

In the summer plots, there is little to choose, TRNSYS being closer but too highly damped. In fact, the reduction of thermal mass makes the TRNSYS fit fairly satisfactorily. As in the winter houses, the ZSTEP shape is good, but offset downwards, with a thermal lag which is too low.

CONCLUSION

The comparison of the two programs with experimental data shows that, as might be expected, the more complex ZSTEP program gives a closer approximation in most cases. Whether the fit with TRNSYS can be said to be close enough to be able to use it where other procedures need to be tested is not immediately obvious. The construction of TRNSYS enables, for instance, the effect of attic fans to be studied. A subroutine in TRNSYS which turns a large slow venting fan on whenever the outside temperature is 1°C or more, lower than internal temperature and indoor temperature is below 20°C, indicates that Hedland in March could be kept comfortable most of the time without air conditioning.

Thus, it would seem a pity to lose the versatility of TRNSYS. This latter program, which applies the hourly heat flow to an assumed building thermal mass, is susceptible to this mass. It would be tempting to assume that the subtracting of the blockwork walls from the thermal mass, giving a better fit, had a real physical significance, but this is open to conjecture.

The behaviour of the programs in the examples chosen is not sufficiently consistent to enable firm decisions on their use to be made.

REFERENCES

(1) Mitalas, G.P. and Arsenault, J.G. (July 1972) Fortran IV Program to Calculate Z-Transfer Functions for the Calculation of Transient Heat Flows Through Walls and Roofs, Division of Building Research, National Research Council of Canada, Ottawa.

(2) Roy, G.G. et al (1981) Procedures to Generate and Edit Data Files for the Program ZSTEP, Research Report No. 9, School of Architecture, University of Western Australia.

Table 1

BRICK VENEER DUPLEX - GERALDTON (DECEMBER)

INSULATED CEILING: HOUSE 'A'

DAY 7/12/79	EXTERNAL DRY BULB TEMP.				TRNSYS				EXPERIMENT				ZSTEP				LAG ON MAXS.			DECREMENT		
	MIN	MAX	TIME OF MIN	TIME OF MAX	MIN	MAX	TIME OF MIN	TIME OF MAX	MIN	MAX	TIME OF MIN	TIME OF MAX	MIN	MAX	TIME OF MIN	TIME OF MAX	EX	TRN	ZST	EX	TRN	ZST
341	13.3	24.4	5	12	21.1	23.0	9	19.5	21.1	27.7	5	16	17.7	24.3	6	16	4	7.5	4	.595	.171	1.04
342	11.4	27.9	5	13	20.3	23.4	7	21.5	20.7	28.7	5	16	16.9	25.9	6	18	3	7.5	5	.485	.188	.812
343	13.7	28.1	5	13.5	22.0	24.8	7	20	21.8	29.5	5	15	18.4	26.7	6	15.5	2.5	6.5	2	.535	.194	.889
344	17.8	29.4	5	10	23.7	26.3	7	19	23.1	30.5	6	16	20.4	27.6	5.5	15	6	9	5	.641	.222	.966
345	17.6	25.2	5	13	24.6	26.0	9	18	23.9	28.1	6	14	21.7	26.0	6	14	1	5	1	.553	.184	1.12
346	13.2	23.3	6	12	22.1	24.4	9	18	21.1	27.0	5	16	18.4	25.1	6	15.5	4	6	3.5	.584	.228	1.13
347	12.2	25.8	4.5	13	20.7	23.1	7	20	20.6	27.9	5	17	17.8	25.8	6	16	4	7	3	.537	.176	.911
348	15.8	24.2	4	11	21.7	23.2	7	19	20.7	27.3	6	17	19.1	25.3	5	16	6	8	5	.631	.179	1.23
349	14.6	23.8	4	13	21.4	23.1	7	20	21.5	27.2	5	16	18.7	25.0	5	15.5	3	7	2.5	.620	.185	1.23
350	12.3	32.7	5	12	21.2	25.6	7	21	20.6	30.3	5	17	18.0	28.7	5.5	16	5	9	6	.475	.216	.725
351	15.4	36.5	4	10	24.3	27.7	6	19	23.2	31.0	5.5	17	19.9	29.1	5	16	7	9	6	.370	.161	.601
352	19.8	24.7	3.5	13	25.1	27.0	11	16.5	24.0	27.5	7	15	22.0	26.0	6	15	2	3.5	2	.714	.387	1.79

UNINSULATED CEILING: HOUSE 'B'

DAY 7/12/79	EXTERNAL DRY BULB TEMP.				TRNSYS				EXPERIMENT				ZSTEP				LAG ON MAXS.			DECREMENT		
	MIN	MAX	TIME OF MIN	TIME OF MAX	MIN	MAX	TIME OF MIN	TIME OF MAX	MIN	MAX	TIME OF MIN	TIME OF MAX	MIN	MAX	TIME OF MIN	TIME OF MAX	EX	TRN	ZST	EX	TRN	ZST
341	13.3	24.4	5	12	17.7	21.9	6	19.5	16.6	27.5	4	16	17.2	26.6	6	14	4	7.5	2	.981	.378	.847
342	11.4	27.9	5	13	19.0	24.0	7	20	18.7	29.1	5	16	16.2	28.4	6	14.5	3	7	1.5	.545	.303	.739
343	13.7	28.1	5	13.5	21.4	25.7	7	19	20.5	30.2	5	16	17.8	29.4	5	14	2.5	5.5	0.5	.604	.298	.805
344	17.8	29.4	5	10	23.4	27.3	7	18	22.3	31.3	5	16	20.1	30.4	5	14	6	8	4	.667	.336	.888
345	17.6	25.2	5	13	24.6	25.1	9	16.5	24.0	27.6	5.5	14	21.6	28.3	5.5	14	1	3.5	1	.645	.197	.882
346	13.2	23.3	6	12	20.9	23.8	8	18	19.7	27.9	5	16	17.8	27.5	6	15	4	6	3	.683	.287	.960
347	12.2	25.8	4.5	13	19.8	23.8	8	19	19.2	28.9	5	16	17.1	28.3	6	14.5	3	6	1.5	.596	.294	.824
348	15.8	24.2	4	11	21.1	23.9	7	18	20.5	27.8	5	16	18.7	27.8	5	15	5	7	4	.714	.333	1.08
349	14.6	23.8	4	13	20.8	23.7	7	18.5	20.1	27.0	5	16	18.2	27.7	5	14	3	6.5	1	.685	.315	1.03
350	12.3	32.7	5	12	20.5	26.8	6.5	19.5	19.2	31.4	5	16	17.3	31.7	5	14	4	7.5	2	.529	.308	.705
351	15.4	36.5	4	10	24.0	28.9	6	18	21.8	32.9	5	16.5	19.4	31.8	5	13	6.5	8	3	.445	.232	.588
352	19.8	24.7	3.5	13	24.5	25.4	10	16	23.5	28.5	7	14.5	21.8	27.8	5.5	15	1.5	3	2	.816	.122	1.22

Table 2

BLOCK VENEER HOUSE - SOUTH HEDLAND (JULY)

INSULATED CEILING: HOUSE 'A'

DAY START: 30/7/79	EXTERNAL DRY BULB TEMP.				TRNSYS				EXPERIMENT				ZSTEP				LAG ON MAXS.			DECREMENT		
	MIN	MAX	TIME OF MIN	TIME OF MAX	MIN	MAX	TIME OF MIN	TIME OF MAX	MIN	MAX	TIME OF MIN	TIME OF MAX	MIN	MAX	TIME OF MIN	TIME OF MAX	EX	TRN	ZST	EX	TRN	ZST
211	7.3	25.3	6	13	18.2	20.8	8	18.5	18.4	24.6	6.5	18	15.9	21.7	6	15.5	5	5.5	2.5	.347	.148	.322
212	7.7	26.6	7	14	18.0	21.3	8	19	19.0	24.9	7	16.5	16.0	22.4	6.5	16	2.5	5	2	.313	.172	.332
213	8.5	24.2	7	12	18.7	21.1	7	20	19.2	25.5	7	16	16.3	22.0	6	15	4	8	3	.402	.151	.363
214	14.5	23.5	2	12	20.0	21.5	6	17.5	21.8	25.1	6	14.5	18.4	22.0	6	15	2.5	5.5	3	.377	.168	.401
215	17.1	23.9	7	16	20.4	22.3	8	21	22.2	25.4	6	16.5	19.1	22.5	6.5	16	.5	5	0	.461	.283	.496
216	14.4	23.6	7	14	21.2	22.4	8	17.5	21.7	25.2	6.5	16.5	18.9	22.6	6	15	2.5	3.5	1	.389	.134	.415
217	14.8	24.0	7	16	20.0	21.9	9	18	21.1	25.3	7	16.5	18.9	22.7	7	16	.5	2	0	.457	.205	.404
218	8.7	23.8	7	15	18.7	20.7	8	18.5	19.5	24.1	7	16.5	17.0	22.1	7	15.5	1.5	3.5	.5	.332	.135	.336
219	8.8	25.2	7	16	18.5	21.3	8	19	19.5	25.5	7	17	16.8	22.5	6.5	16.5	1	3	.5	.370	.172	.340
220	12.0	26.0	3	16	20.0	22.6	7.5	18	19.7	25.5	4	17	18.3	23.2	2	16.5	1	2	.5	.412	.191	.344
221	9.2	26.9	7	16	19.5	22.7	8	19	19.5	25.9	7	17	17.3	23.1	7.5	17.5	1	3	1.5	.362	.176	.328
222	12.6	27.2	2	17	20.6	23.7	7	19	20.6	26.2	6	17	18.4	23.7	4	18	0	2	1	.386	.212	.362

UNINSULATED CEILING: HOUSE 'B'

DAY START: 30/7/79	EXT. MIN	EXT. MAX	EXT. TIME OF MIN	EXT. TIME OF MAX	TRNSYS MIN	TRNSYS MAX	TRNSYS TIME OF MIN	TRNSYS TIME OF MAX	EXPERIMENT MIN	EXPERIMENT MAX	EXPERIMENT TIME OF MIN	EXPERIMENT TIME OF MAX	ZSTEP MIN	ZSTEP MAX	ZSTEP TIME OF MIN	ZSTEP TIME OF MAX	LAG EX	LAG TRN	LAG ZST	DECREMENT EX	DECREMENT TRN	DECREMENT ZST
211	7.3	25.3	6	13																		
212	7.7	26.6	6	13	13.9	24.0	7	17	17.3	25.7	7	17	16.8	25.0	6	13	4	4	0	.443	.533	.433
213	8.5	24.2	6	11	14.4	23.2	7	17	17.5	25.8	7	17	17.1	24.3	6	13	6	6	1	.523	.558	.463
214	14.5	23.5	1	11	17.6	22.7	7	16	20.4	24.9	7.5	15.5	19.2	23.4	2.5	14	3.4	4	2	.509	.569	.457
215	17.1	23.9	6	15	18.1	24.1	7	17	20.7	25.5	7.5	17.5	19.7	24.1	6	14	2.5	2	0	.698	.876	.628
216	14.4	23.6	6	13	18.4	24.4	7	16	20.4	25.6	7.5	17	19.3	24.6	6	13	4	3	0	.574	.658	.591
217	14.8	24.0	6	15	17.7	23.8	7	17	19.7	25.6	8	17.5	19.5	24.7	6.5	15	2.5	2	0	.649	.667	.526
218	8.7	23.8	6	14	14.7	22.8	7	17	18.3	25.3	8	17.5	17.6	24.4	6.5	14	3.5	3	0	.459	.535	.436
219	8.8	25.2	6	15	14.4	23.7	7	17.5	18.2	25.6	7.5	17	17.5	24.5	6.5	14.5	2	2.5	0	.449	.565	.419
220	12.0	28.0	2	14	17.5	25.6	7	17	18.1	26.1	8	18	18.7	25.0	2	15	4	3	1	.572	.581	.430
221	9.2	26.9	6	15	15.4	25.0	7	17	18.3	26.1	8	18	17.8	24.5	7.5	16	3	2	0	.438	.542	.378
222	12.6	27.2	3	16	17.5	26.5	6	17	18.9	26.4	7.5	17.5	18.9	25.0	4.5	15.5	1.5	1	0	.512	.615	.417

Table 3

BLOCK VENEER HOUSE - SOUTH HEDLAND (FEBRUARY)

INSULATED CEILING: HOUSE 'A'

DAY START: 24/2/80	EXTERNAL DRY BULB TEMP.				TRNSYS				EXPERIMENT				ZSTEP				LAG ON MAXS.			DECREMENT		
	MIN	MAX	TIME OF MIN	TIME OF MAX	MIN	MAX	TIME OF MIN	TIME OF MAX	MIN	MAX	TIME OF MIN	TIME OF MAX	MIN	MAX	TIME OF MIN	TIME OF MAX	EX	TRN	ZST	EX	TRN	ZST
54	23.6	37	5	12	29.4	32.2	7	19	29.7	34.6	5	17	28.9	33.6	6	15.5	5	7	3.5	.366	.209	.276
55	23.9	32.9	4	14	29.9	31.7	9	18	30.6	34.2	7	16.5	29.0	32.7	7	15.5	2.5	4	1.5	.404	.202	.416
56	25.2	39.4	5	13	29.8	33.0	7	19	30.8	35.4	5.5	17	28.9	33.8	6	15.5	4	6	2.5	.323	.225	.275
57	24.7	36.0	4	13	31.5	33.5	7	17	31.3	35.4	6	16	29.3	33.7	6	15.5	3	4	2.5	.363	.177	.389
58	26.1	32.9	5	13	31.5	32.8	8	18	31.9	35.0	6	16	29.5	32.9	6	15.5	3	5	2.5	.456	.191	.500
59	25.7	36.7	5	14	30.8	32.9	8	18.5	31.7	35.8	6.5	15	29.2	33.7	6	15.5	2	4.5	1.5	.373	.191	.409
60	26.4	39.0	5	12	31.2	33.9	7	18	32.1	35.8	6.5	16.5	29.7	34.0	7	11.4	4.5	6	2	.296	.216	.344
61	26.5	35.7	5	13	31.7	33.2	7	18	32.0	36.0	6.5	15	30.0	33.7	6.5	15	2	5	2	.435	.163	.402
62	26.4	37.3	5	12	31.6	33.6	8	17.5	32.3	36.3	6	16	29.8	33.9	7	15	4	5.5	3	.367	.183	.376

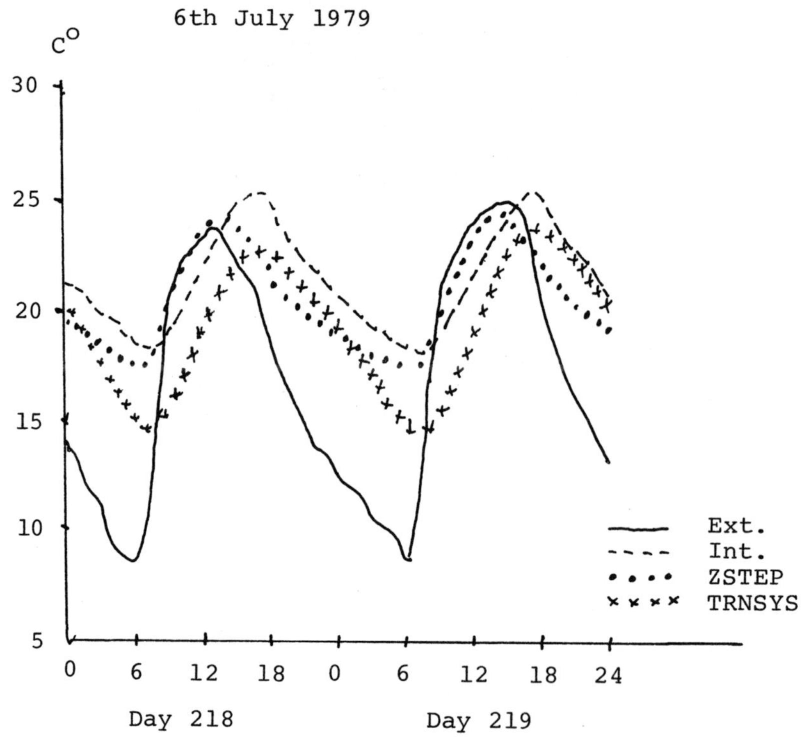

Fig. la Port Hedland winter uninsulated house

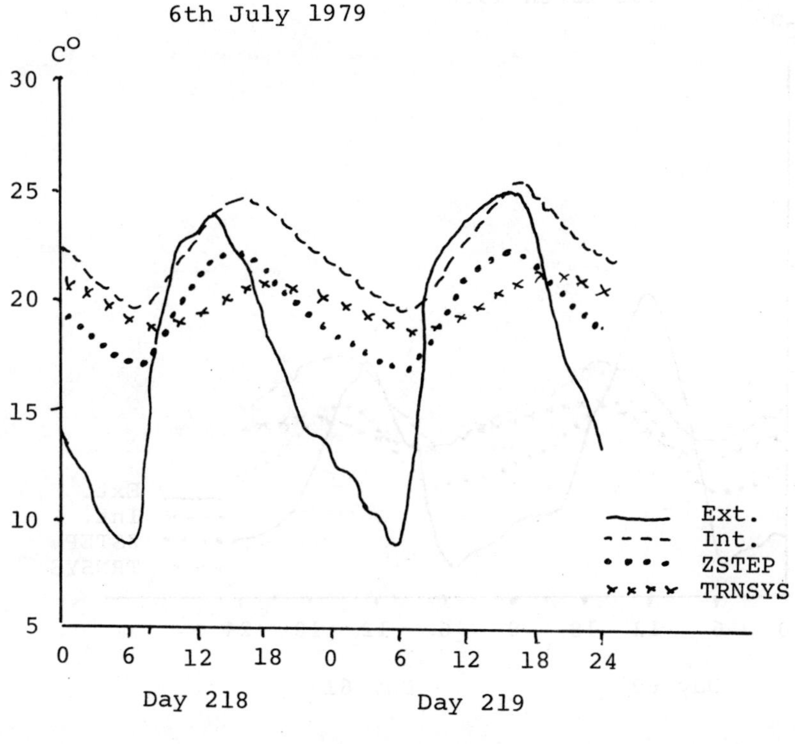

Fig. 1b Port Hedland winter insulated house

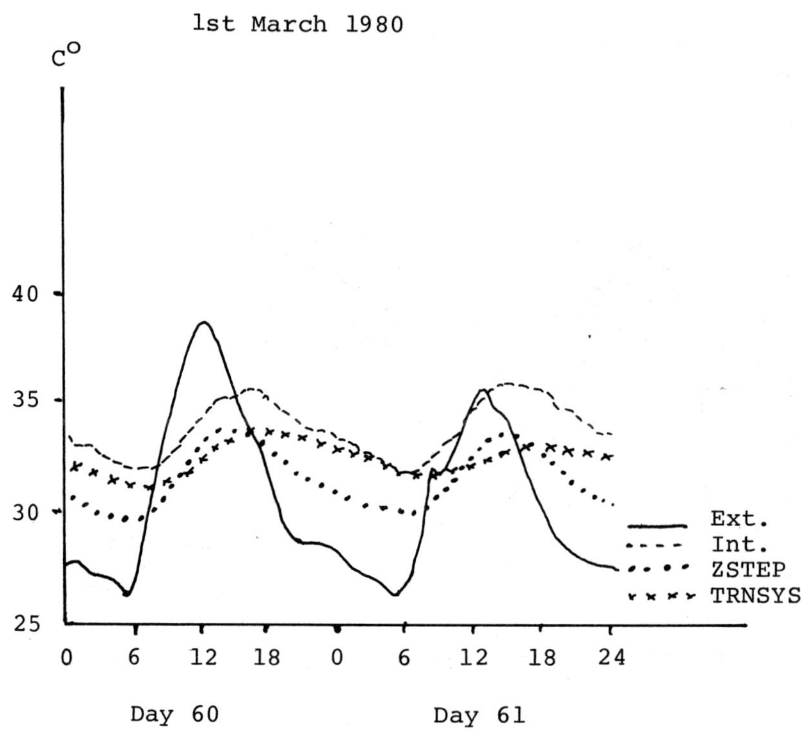

Fig. 2 Port Hedland summer insulated

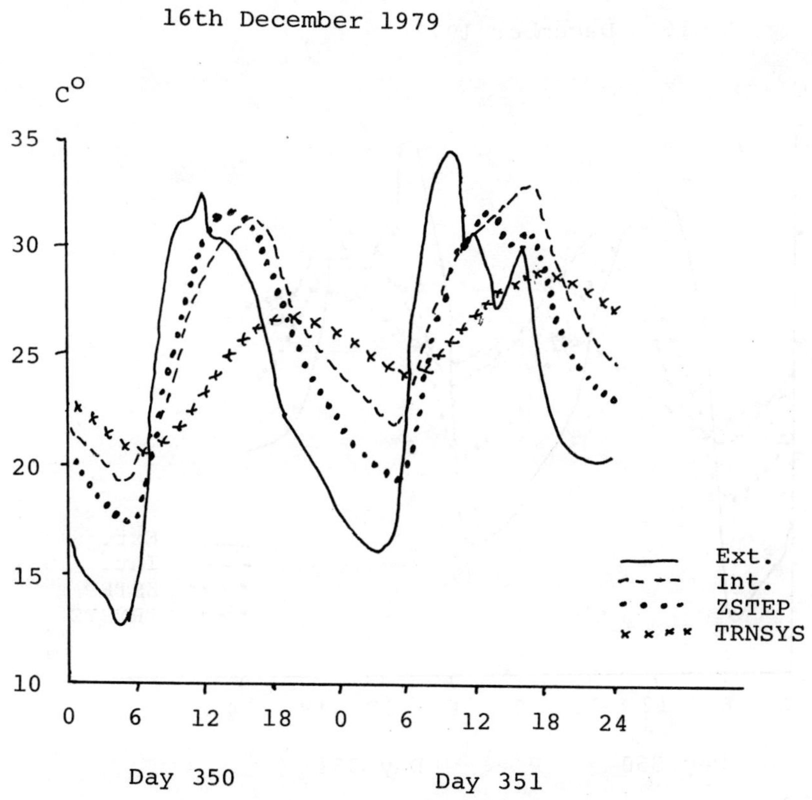

Fig. 3a Geraldton summer uninsulated house

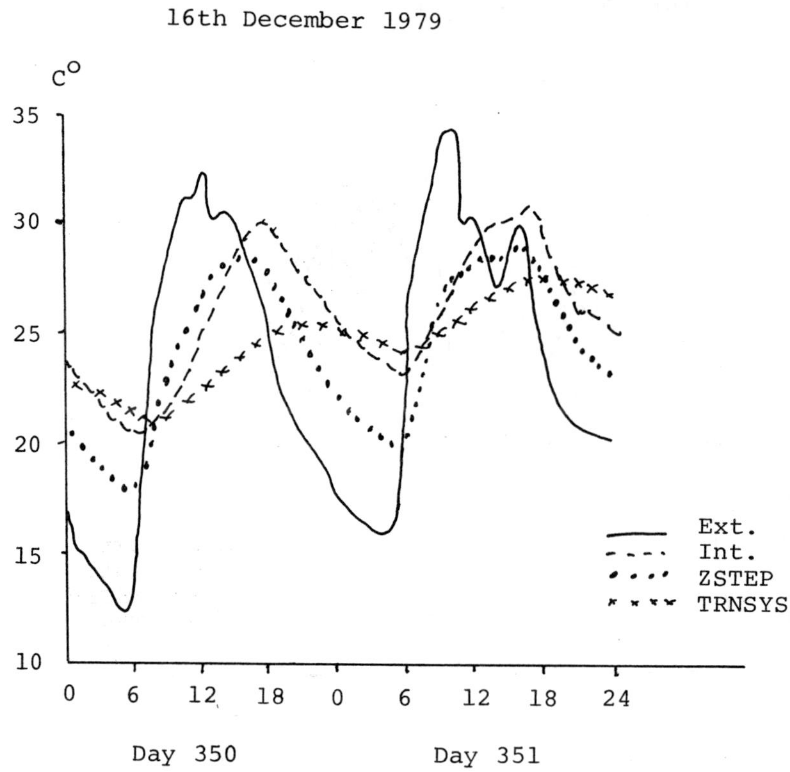

Fig. 3b Geraldton summer insulated house

An Empirical Study of Problem Heat Flow Paths in Simulation Models

M. Smart and J. Ballinger
Solarch, Solar Architecture Research Unit,
University of New South Wales

ABSTRACT

Multiple Linear regression analysis is applied to temperature measurements inside fifteen unconditioned unoccupied houses.

It is shown that average building performance may be described to within measurement tolerances by a set of three coefficients.

The possible use of these coefficients for validation of simulation models and as a basis for simplified prediction methods is discussed.

I INTRODUCTION

Twelve passive solar houses and three conventional housing commission houses were built in a single development at Bonnyrigg in Sydney's western suburbs in 1981. Temperatures, power consumption figures and weather data are recorded on an hourly basis at a central monitoring station. This paper will describe measurements taken between February and March 1982. During this time the fifteen houses remained unoccupied and unconditioned.

The Bonnyrigg Solar Village provides a unique opportunity to study in detail the thermal performance of a variety of house designs under controlled conditions. Comparisons between prediction and measurement have previously been hampered by a lack of well-documented measurements. Monitored houses have frequently been at isolated sites, hence inter-comparisons between houses have been complicated by climatic differences.

In presenting the monitoring results from Bonnyrigg we will attempt to keep the analysis as free as possible from any particular models or theoretical bias. However, there is an enormous quantity of information. Any attempt to present it requires simplifications of some kind, which inevitably lead to models. Therefore we adopt, as a basis for presentation, a linear model: the daily averaged interior condition is assumed to be a linear function of daily averaged insolation, ambient air temperature, and ground temperature. The daily fluctuation of interior conditions is assumed to be a linear function of daily fluctuations in weather conditions.

The motivation for such a simplified approach is the need to come to grips with certain 'problem heat flow paths'; aspects of simulation modelling which are difficult to account for using pure thought and for which relevant measurements either are too difficult to conduct or have not been made.

Specifically, estimating the net total solar gain of a structure is a complex task involving a calculation of the net absorption of reflected, diffuse, and direct radiation by opaque and transparent surfaces (Fig. 1). Quantifying the thermal resistivity of the building envelope is difficult in theory because of the wind dependent surface air film coefficients and ventilation rate. The influence of the ground is highly dependent upon the unknown interface between slab floor and ground. Justly or not in the case of suspended floors the ground influence is often assumed to be neglible.

The measurements are described in detail in Section II. A linear model for temperature averages and fluctuations is proposed and the regression analysis of this model explained in Section III. The regression results are discussed in Section IV. Possible extensions of the analysis and other applications are considered in Section V.

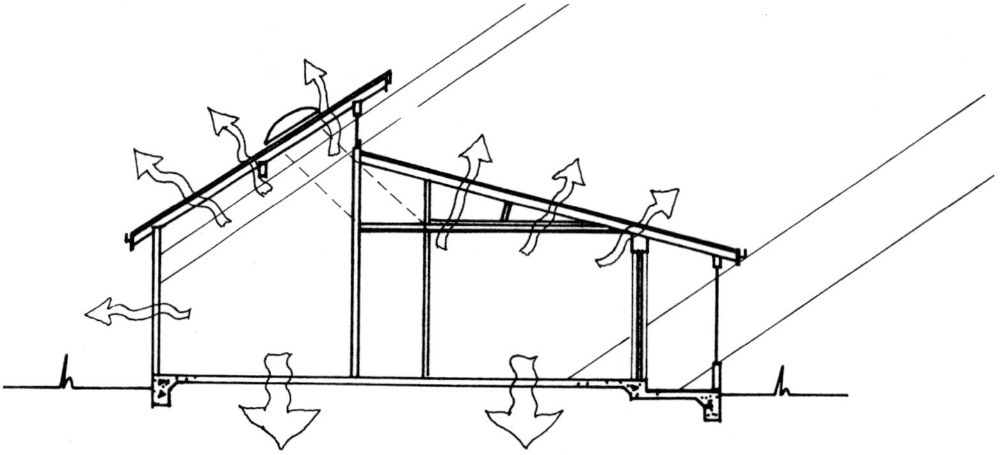

Fig. 1 Problem areas in estimating energy balance: overall solar gain, conductive and convective losses through the building envelope, and conduction into the ground.

II MEASUREMENTS

A perspective drawing of the Solar Village (Fig. 2) shows the monitoring station and lengths of underground cable connecting it to each house. Houses will be referred to by their three-digit lot numbers: 340-351 are the solar designs and 357-359 are the standard houses.

Data are recorded automatically by a Hewlett-Packard 3054A data acquisition system which has been configured to sample type T thermocouples at one hour intervals and by a Digital Electronics EDAS 16B configured to accumulate multichannel pulsed input at ten minute intervals. Pyranometers are connected to the EDAS 16B via voltage-to-frequency converters.

Provision has been made to measure detailed electricity consumption in each house by means of six and ten pair telephone cables connecting the power meters to the logging equipment. Revolutions of the disc in the meter are converted to pulses by an opto-isolator. However no power was used and no heating or air-conditioning was applied during the test period: February 5 - March 20.

A weather station is mounted on the roof of the monitoring station. It consists of a Rimco 4 inch cup anemometer and a Rimco wind direction unit on a nine metre tower, two Eppley pyranometers, one of which has a shade ring, and a shielded, aspirated thermocouple to measure ambient dry bulb temperature. The pyranometers measure Global and diffuse radiation on a horizontal surface. The error tolerance for each of these is less than one per cent at full scale, or 10 Watts per metre squared. The thermocouple is mounted under a white metal hood attached to the shaded south side of the monitoring building. Air is drawn over the thermocouple at the fixed rate of 60 metres per second by a small fan. According to the ANSI standard C96.1 the limits of error for type-T thermocouples in the temperature range -60 to 100 degrees C are +/- .56-.28 degrees C. (1).

A total of one hundred thermocouple points are distributed among the fifteen houses. The number of temperature probes allocated to each house varies, however the living room temperature in each house, the slab temperature in the ten concrete floor houses, and the subfloor space temperature in the five suspended floor houses are measured. Ground temperatures are measured at depths of 300mm and 600mm in the concrete floor houses and also at 1m in house #346. Many of the houses contain probes in a south-facing bedroom in addition to the north-facing living room.

A test for spatial variability of internal air temperatures was conducted in three of the houses over two and four day periods. Variations on the order of +/- 1 degree C were found at some times of day, however the daily averaged discrepancies were much smaller. The location of probes for this test is illustrated in Fig. 3. Fig. 4 also illustrates the placement of ground probes.

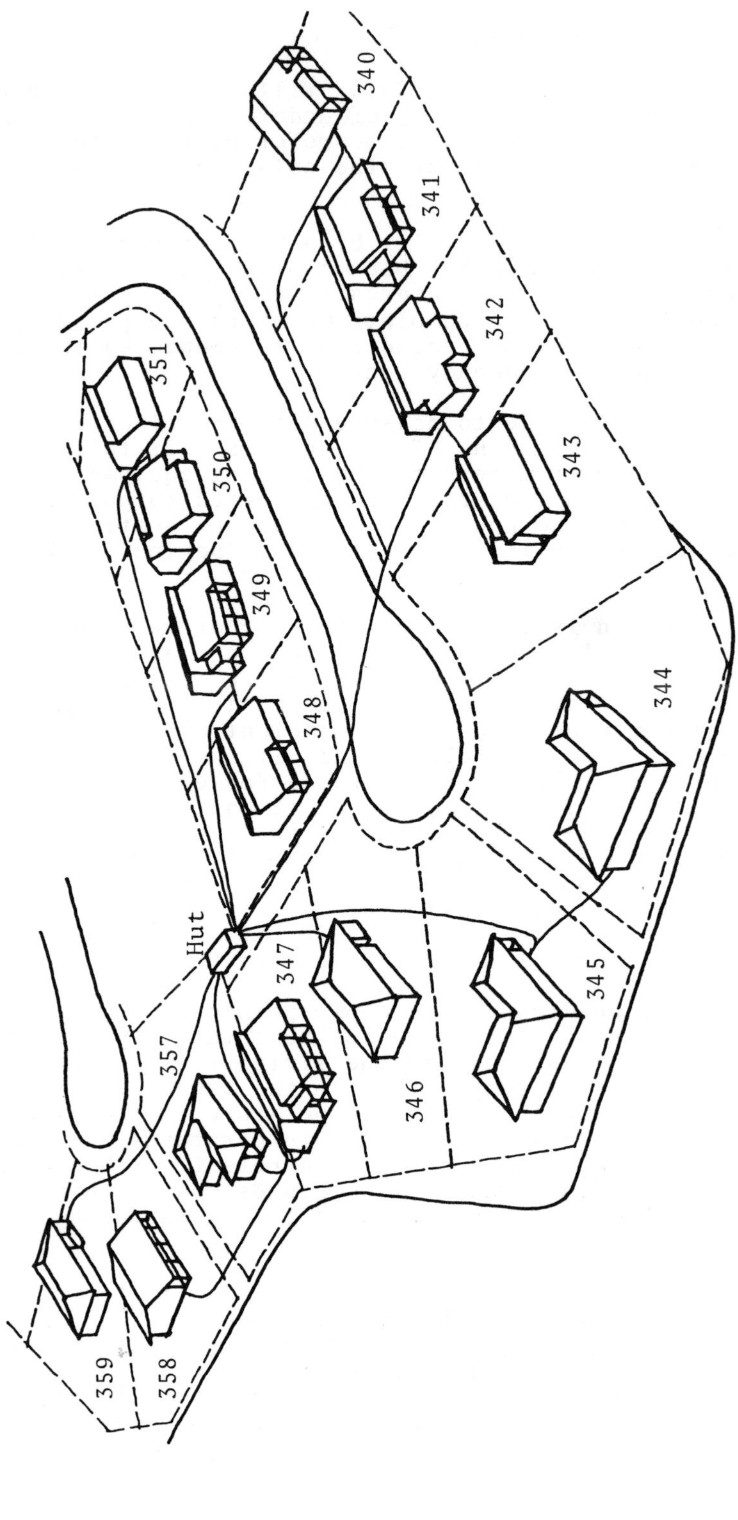

Fig. 2 Perspective of the Bonnyrigg site.

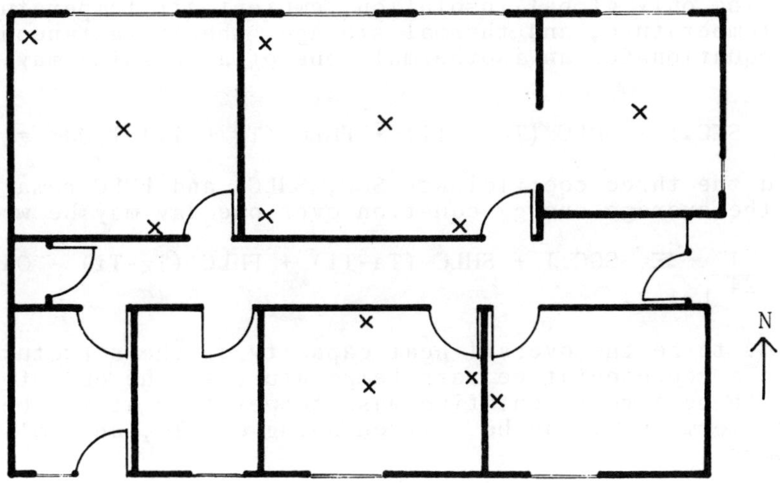

Fig. 3 Plan of house #343 showing probe locations during short-term test.

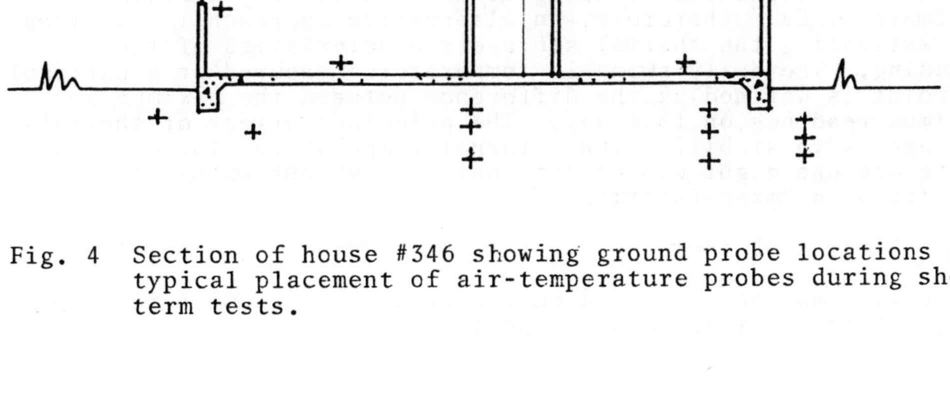

Fig. 4 Section of house #346 showing ground probe locations and typical placement of air-temperature probes during short term tests.

III A. THE ENERGY EQUATION

Considering only global insolation, ambient air temperature, ground temperature, and thermal storage, the instantaneous energy equation for an isothermal zone of a building may be written

$$SGC.I + SHLC\ (Ta - Ti) + FHLC\ (Tg - Ti) - Qam = 0 \quad (I.)$$

Provided the three coefficients SGC, SHLC, and FHLC remain constant, the average energy equation over one day may be written

$$\frac{1}{24} \sum_{h=1,24} SGC.I + SHLC\ (Ta-Ti) + FHLC\ (Tg-Ti) - Qam=0 \quad (II)$$

Taking Cv to be the overall heat capacity of the structural mass, Tm to be a representative mass temperature at the end of the day, and Tm' to be a representative mass temperature at the beginning, the last term in II may be written using Cv, Tm, and Tm'. Thus II becomes

$$SGC.\overline{I} + SHLC\ (\overline{Ta}-\overline{Ti}) + FHLC\ (\overline{Tg}-\overline{Ti}) - \frac{Cv}{24}\ (Tm-Tm')=0 \quad (III.)$$

The temperatures and radiation figures in III are all directly measureable quantities. The four coefficients SGC, SHLC, FHLC, and Cv are characteristics of the building in question and, in principle, may be deduced by regression on the temperature and insolation measurement.

In fact the assumption of constancy for the four coefficients is not strictly true. The solar gain coefficient is highly dependent upon the sun's position. Passive solar houses are specifically designed to admit the noon sunlight in winter and to obstruct it in summer. The shell heat loss coefficient is heavily influenced by the wind-dependent air-film layers and ventilation rate. To a lesser extent, the floor heat loss coefficient is affected by soil moisture content.

Furthermore the difference between starting and ending mass temperatures is likely to be very small compared to the other terms in III, possibly leading to a comparatively unreliable estimate of Cv. Therefore, an alternative approach is required for estimating the thermal storage characteristics of the building. The daily swing in temperature measured at a particular point is defined as the difference between the maximum and minimum readings on that day. The principal effect of thermal storage is to stabilize the internal temperature fluctuations. Therefore one might expect internal and ambient swings to be related by a "mass function":

$$Si = MF\ (Sa) \quad (IV.)$$

Regression may be performed on the set of measurement pairs (Sa, Si) to determine a mass function.

III B. REGRESSION ANALYSIS OF MEAN TEMPERATURES

To solve III for the coefficients it is necessary to rewrite the equation in terms of ratios of the coefficients. Thus new coefficients Pi, i=1....4 are introduced:

$$P1 = \frac{SGC}{SHLC + FHLC}$$

$$P2 = \frac{SHLC}{SHLC + FHLC}$$

$$P3 = \frac{FHLC}{SHLC + FHLC}$$

$$P4 = \frac{Cv}{24 \times (SHLC + FHLC)}$$

$$\overline{Ti} = P1\ \overline{I} + P2\ \overline{Ta} + P3\ \overline{Tg} + P4\ (Tm - Tm') \qquad (V.)$$

Solution of V will yield ratios of building coefficients but not actual values. The problem is not simply a mathematical one. Temperature and insolation measurements lack any reference to spatial dimension. Therefore any coefficients derived from these measurements must also lack reference to a spatial dimension.

To give a crude example, it is conceivable that a small building would have comparable internal temperatures to a large one in the same area, and that the relative influences of insolation, air temperature, ground temperature, and thermal mass, as described by the Pi, are also comparable. However the actual heat flows would scale to the size of the building and the solar gain, shell heat loss, and floor heat loss coefficients would all be larger in the larger building. Therefore actual values for these coefficients cannot be established without some reference to a spatial dimension.

Such a reference is available in a conditioned building, however. The energy used for heating would appear on the left-hand side of III with a negative sign. Since the units would be energy and the insolation units are energy per unit area, a spatial reference must be included to reconcile the two types of measurement. This point will be discussed further in the recommendations section.

Equation V is a linear equation in four independent variables of the type:

$$Yk = P1\ X1k + P2\ X2k + P3\ X3k + P4\ X4k$$

where we have one set of measurements (X1,X2,X3,X4,Y) each day (k). Because of the spatial variability within each house measurements of Ti, are subject to the greatest variance about a truly representative value.

The Xik are comparatively accurate measured values. The choice of a set of Pi makes it possible to estimate a value for Yk. Call the estimate $\hat{Y}k$. Given n days of measurement we wish to choose a set of Pi for which the following is minimized.

$$\sum_{k=1}^{n} \left(Yk - \hat{Y}k \right)^2$$

Differentiating the sum of squares difference with respect to each Pi and setting each partial differential equal to zero yields four 'normal equations'. These may be written in matrix form. (All summations are presumed to go from k=1 to n. Subscripts k are dropped for brevity).

$$\begin{bmatrix} \Sigma X1^2 & \Sigma X1.X2 & \Sigma X1.X3 & \Sigma X1.X4 \\ \Sigma X1.X2 & \Sigma X2^2 & \Sigma X2.X3 & \Sigma X2.X4 \\ \Sigma X1.X3 & \Sigma X2.X3 & \Sigma X3^2 & \Sigma X3.X4 \\ \Sigma X1.X4 & \Sigma X2.X4 & \Sigma X3.X4 & \Sigma X4^2 \end{bmatrix} \begin{matrix} P1 \\ P2 \\ P3 \\ P4 \end{matrix} = \begin{matrix} \Sigma X1.Y \\ \Sigma X2.Y \\ \Sigma X3.Y \\ \Sigma X4.Y \end{matrix} \qquad \text{(VI.)}$$

Multiplying both sides of VI by the inverse matrix we obtain expressions for the coefficients Pi.

Finally to determine how good an estimate for the coefficients has been obtained we divide the sum of squares difference by n-1. The figure thus calculated is the residual variance of measured values of Ti about the least squares estimate of Ti. If the variance is consistent with the known measurement tolerance for Ti then the expression V is a credible description of the mean thermal performance. However, if the variance exceeds measurement tolerances then the hypothetical relationship V must be viewed with some skepticism.

III C. REGRESSION ANALYSIS OF TEMPERATURE SWINGS

Fig. 5 is a plot of daily swings in ambient air temperature versus living room air temperature swings for 122 days. The overall form suggests a monotonically increasing function which intersects the horizontal axis near the origin. The general linear nature of building thermal processes suggests that the mass function should be linear, and the obvious scatter of the points makes it seem unlikely that a higher-order function would provice a better fit. For these reasons we apply linear regression again, but the analysis is simpler since there is only one independent variable.

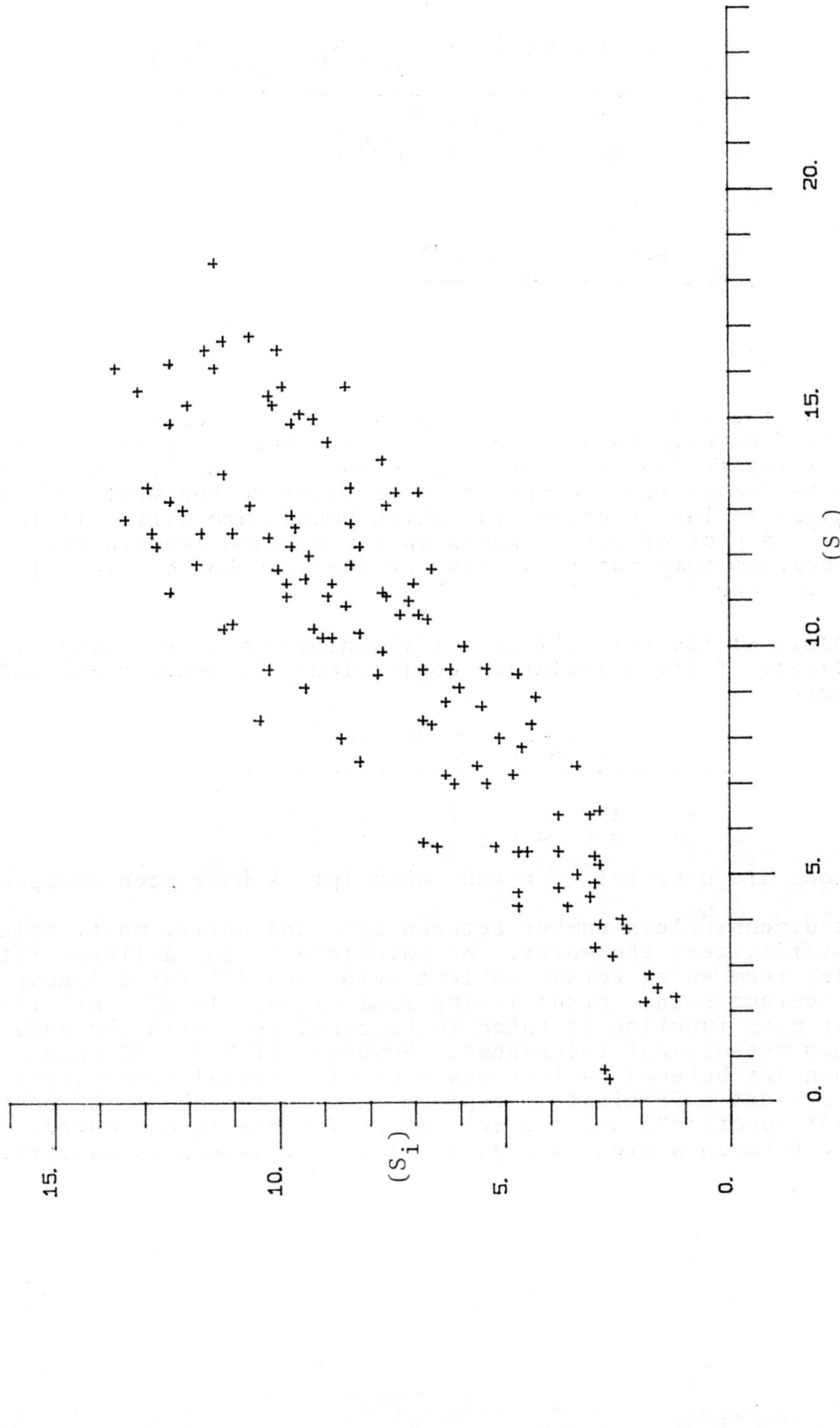

Fig. 5 Plot of ambient air-temperature swing vs. living room temperature for house #357.

A best fit straight line may be estimated once more by the method of least squares. The slope, B, and y-intercept, A are given by

$$B = \frac{\sum_{k=1}^{n}\left(S_{ak}.S_{ik}\right) - \frac{1}{n}\sum_{k=1}^{n}\left(S_{ak}\right)\sum_{k=1}^{n}\left(S_{ik}\right)}{\sum_{k=1}^{n} S_{ak}^{\ 2} - \frac{1}{n}\left(\sum_{k=1}^{n} S_{ak}\right)^2}$$

$$A = \frac{\sum_{k=1}^{n} S_{ik}}{n} - B\,\frac{\sum_{k=1}^{n} S_{ak}}{n}$$

In testing the validity of this approach it is necessary to proceed differently than in Section IIIB. Whereas error tolerances for twenty-four hour average measurements of Ti could possibly be as low as thermocouple error limits, measurements of Si are subject to larger errors for which bounds are difficult to estimate. A plot of daily swings in living room temperature versus bedroom temperature swings for the same house (Fig. 6) illustrates the difficulty.

The quality of the straight line fit achieved may be gauged by the estimate of the correlation coefficient, R, between Sak and Sik, where

$$R^2 = \frac{\left(\Sigma S_a.S_i - \frac{1}{n}\,\Sigma S_a.\Sigma S_i\right)^2}{\left[\Sigma S_a^{\ 2} - \frac{1}{n}\left(\Sigma S_a\right)^2\right]\left[\Sigma S_i^{\ 2} - \frac{1}{n}\left(\Sigma S_i\right)^2\right]}$$

(Sumations are over k=1 to n and subscripts k have been dropped).

R2 is a dimensionless number between zero and unity, unity being the best fit, zero the worst. We calculate R2 for a linear fit to living room swing versus ambient swing and R'2 for a linear fit to bedroom swing versus living room swing. If R2 R'2 then a linear mass function is taken to be consistent with the data to within measurement tolerances. However, if R'2 R2 then discrepancies between various measures of internal temperature swings are not sufficient to explain the scatter about the best-fit "mass function", and the existence of a one-to-one correspondence between ambient swings and internal swings is cast into suspicion.

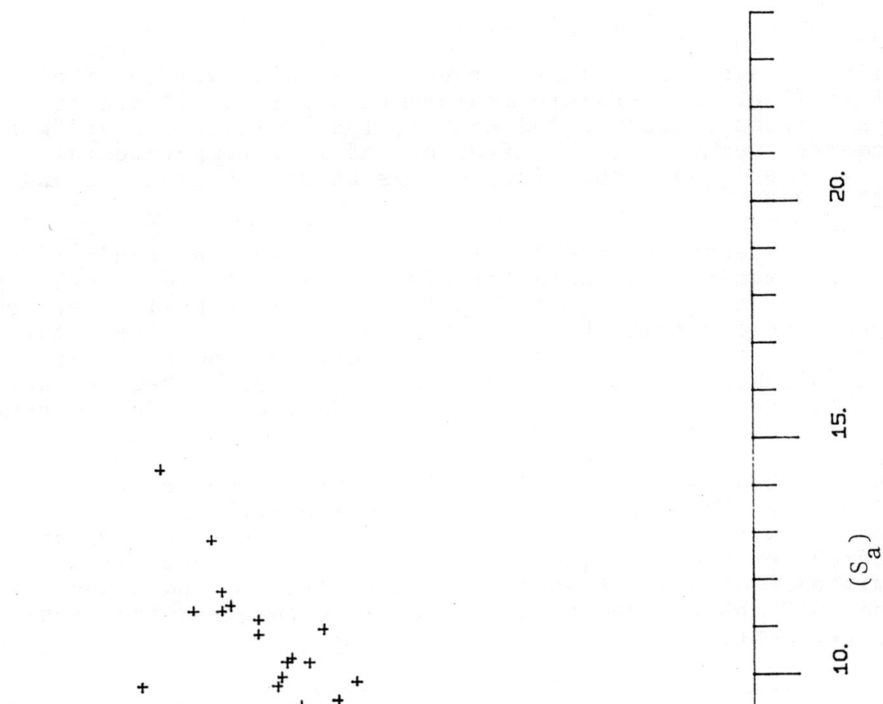

Fig. 6 Plot of bedroom temperature swing vs. living room temperature swing for house #357.

IV. RESULTS AND DISCUSSION

IV A. SELECTION OF DATA

Aware of the danger that data will confess to almost anything if tortured long enough, we have attempted not to select data. We use daily averages and swings for every day available at the time of writing. Temperature measurements were available for 122 days between January 5 and May 11, 1982. Radiation and wind measurements, recorded on a different and more temperamental device, were only available for 40 days between February 5 and March 20, 1982.

The shell heat loss coefficient is sensitive to wind conditions. Although we examined the data for wind variability we eventually decided not to omit any of it from the analysis. Daily averaged wind speeds fell within the range 6.4 - 14.1 km/hr. The solar gain coefficient is sensitive to solar position and hence is likely to change with the time of year. However we decided not to split up the continuous 40 day period during which concurrent measurements of temperature, radiation, and wind were available.

Three of the houses were omitted and two of the houses were studied over a shortened period. #343 was omitted because its thermocouple wiring was damaged by vandalism. #348 was occupied by research personnel during much of the period, hence it was also omitted. Studies of house #358 were omitted and houses #346 and #359 abbreviated because they were involved in a separate experiment.

IV B. DISCUSSION OF MEAN TEMPERATURE REGRESSION RESULTS

Coefficients Pi and the square root o of the residual variance are given for twelve of the fifteen houses in Table 1. The houses with concrete floors on grade exhibit relatively high floor coefficients ($0.6 < P3 \leq 0.8$) whereas the suspended timber floor houses exhibit relatively low floor coefficients ($P3 < 0.6$). The houses with insulated walls and draught-sealed windows and doors have low shell coefficients ($P2 \leq 0.5$) but the uninsulated houses have high shell coefficients ($P2 > 0.7$). The relationship between radiation gain coefficients and orientation or shading of windows is not so clear. One must bear in mind, however, that P1 is the ratio of solar gain coefficient to the sum of shell and floor heat loss coefficients.

Excepting houses #344 and #349, which have unusually high residual variances, σ lies within the range 0.34 - 0.59 degrees C for all houses. This range of errors is consistent with the ANSI standard error limits for T-type thermocouples quoted above: +/- 0.56 - 0.28 degrees C.

Table 1						
House	Description	P1	P2	P3	P4	o
340	slab, insul	2.4e-3	0.33	0.70	-0.64	0.35
341	slab, insul	2.2e-3	0.28	0.80	0.041	0.52
342	slab, insul	1.9e-3	0.38	0.67	-0.007	0.35
344	timber, insul	1.4e-3	0.50	0.58	-0.209	0.81
345	slab, insul	3.5e-4	0.41	0.68	0.241	0.59
346	slab, insul	1.1e-3	0.31	0.75	0.227	0.37
347	slab, insul	5.5e-4	0.30	0.77	0.107	0.49
349	timber, insul	2.4e-3	0.61	0.46	-0.111	1.16
350	slab, insul	2.6e-3	0.42	0.64	-0.047	0.43
351	slab, insul	2.9e-3	0.44	0.61	-0.097	0.34
357	timber, unins	3.1e-3	0.72	0.32	-0.061	0.50
359	timber, unins	1.6e-3	0.84	0.23	-0.028	0.58

An unexpected result of this analysis is a quantifiable influence of ground temperature on suspended floor houses. To test whether this inferred influence is real, we proposed a null hypothesis. We recalculated the residual variance for suspended floor houses under the hypothetical assumption that the floor heat loss coefficient, reflecting a ground influence, is zero and heat loss through the wooden floor is included in the shell heat loss coefficient, reflecting an ambient air influence. In all four cases, σ exceeded 1.7 degrees C, placing the residual variance well beyond the expected measurement error. The null hypothesis loses credibility.

A puzzling aspect of the analysis is that the sum of P2 and P3 exceeds unity. This result is unexpected since P2 was intended to be the ratio of shell to shell-plus-floor heat loss coefficients and P3 the ratio of floor to shell-plus-floor heat loss coefficients. At this stage we can only comment that such a result could arise if there were a systematic overestimate of Ti or underestimate of either Ta or Tg. In this connection it should be mentioned that ground temperature is measured at a depth of 300mm and that systematic discrepancies are likely between this ground temperature and the temperature at the slab-ground interface.

Examination of the sign of derived values for P4 reveals that they are sometimes positive, although the energy equation requires that P4 be negative definite. The effect on σ of neglecting the term P4 (Tm-Tm') is demonstrated in Table 2. Included are six houses with high thermal capacity. The table shows that the contribution of this term is negligible because the change in mass temperature is so slight. Therefore the regression analysis is relatively insensitive to the independent variable (Tm-Tm') and the estimate for P4 is comparatively unreliable. Our initial concern that analysis of mean conditions might give unreliable estimates of structural heat capacity motivated the study of thermal mass by analysis of temperature swings.

Table 2

House	P4	o	o(P4=0)
344	-0.13	0.83	0.86
349	-0.11	1.16	1.18
357	-0.05	0.55	0.58
359	-0.02	0.60	0.61
341	0.04	0.53	0.53
345	0.24	0.59	0.61
346	0.23	0.39	0.43
347	0.11	0.49	0.51

IV C. RESULTS OF ANALYSIS ON TEMPERATURE SWINGS

Slopes of best-fit lines to plots of ambient swing versus living-room swing vary markedly between lightweight construction (Fig. 7) and heavyweight construction (Fig. 8). As table 3 indicates, there is a strong inverse correlation between the amount of masonry or concrete, and the slope of the line. None of the best-fit lines intersect the origin, although the y-intercepts mostly lie within 1 degree C of the origin. Considering the scatter of the points this fact does not seem surprising.

Table 4 quotes the quality of fit indicators R2 for ambient versus living room and living room versus bedroom in those houses which have bedroom probes. In houses #351, 357 the ambient versus living room line is a better fit. However in general it is not fair to assert that the scatter of the data about the proposed mass function can be explained by discrepancies between different measures of internal swing.

A more likely explanation for the scatter is the role of insolation. Particularly in wintertime, the fluctuations which drive temperature changes in thermal storage media have both temperature and insolation components. We discuss a method of taking the insolation component into account in the recommendations section.

Although the proposed linear mass function fails the validity test we have applied, the qualitative nature of the assumption is valid, and the relatively high correlation coefficients obtained provide confidence that this method embodies a useful approach to the problem of thermal mass.

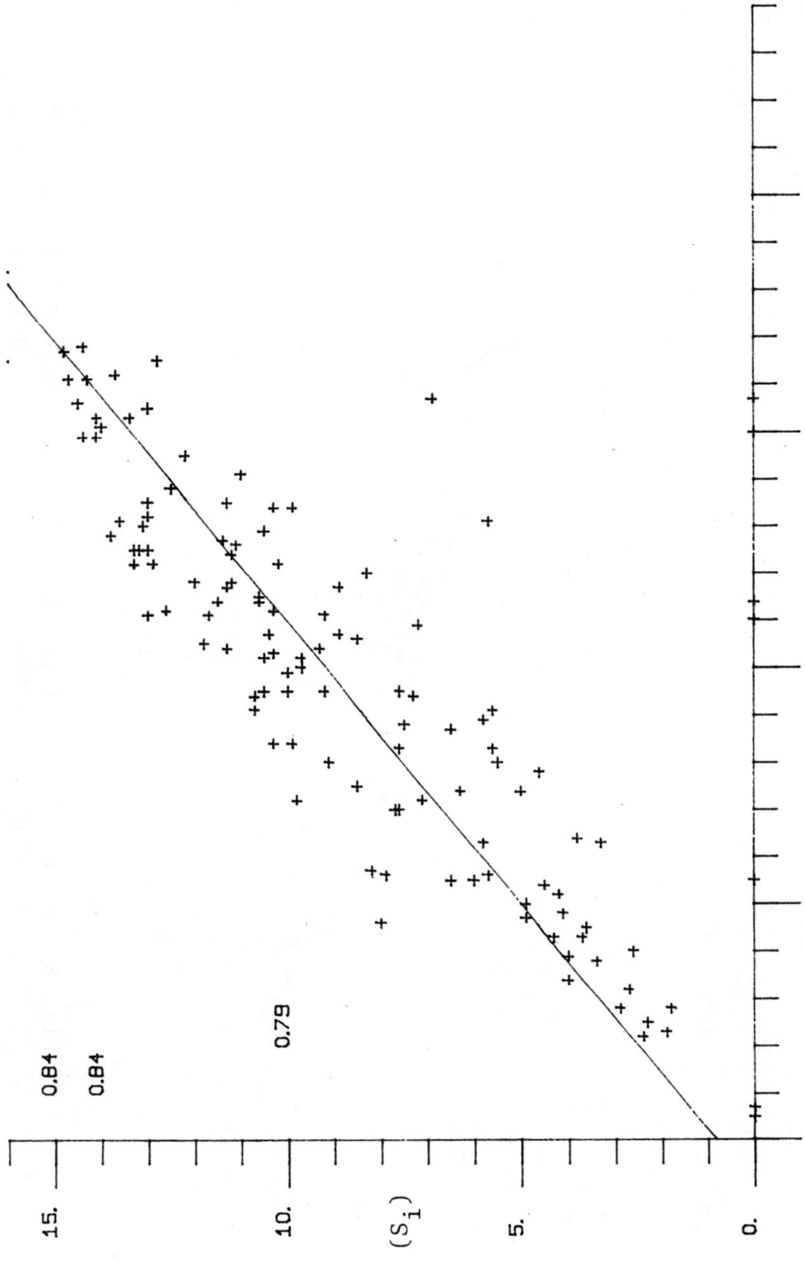

Fig. 7 Plot of ambient vs. living room swings for house #359 showing best linear fit to the data. #359 is a lightweight house.

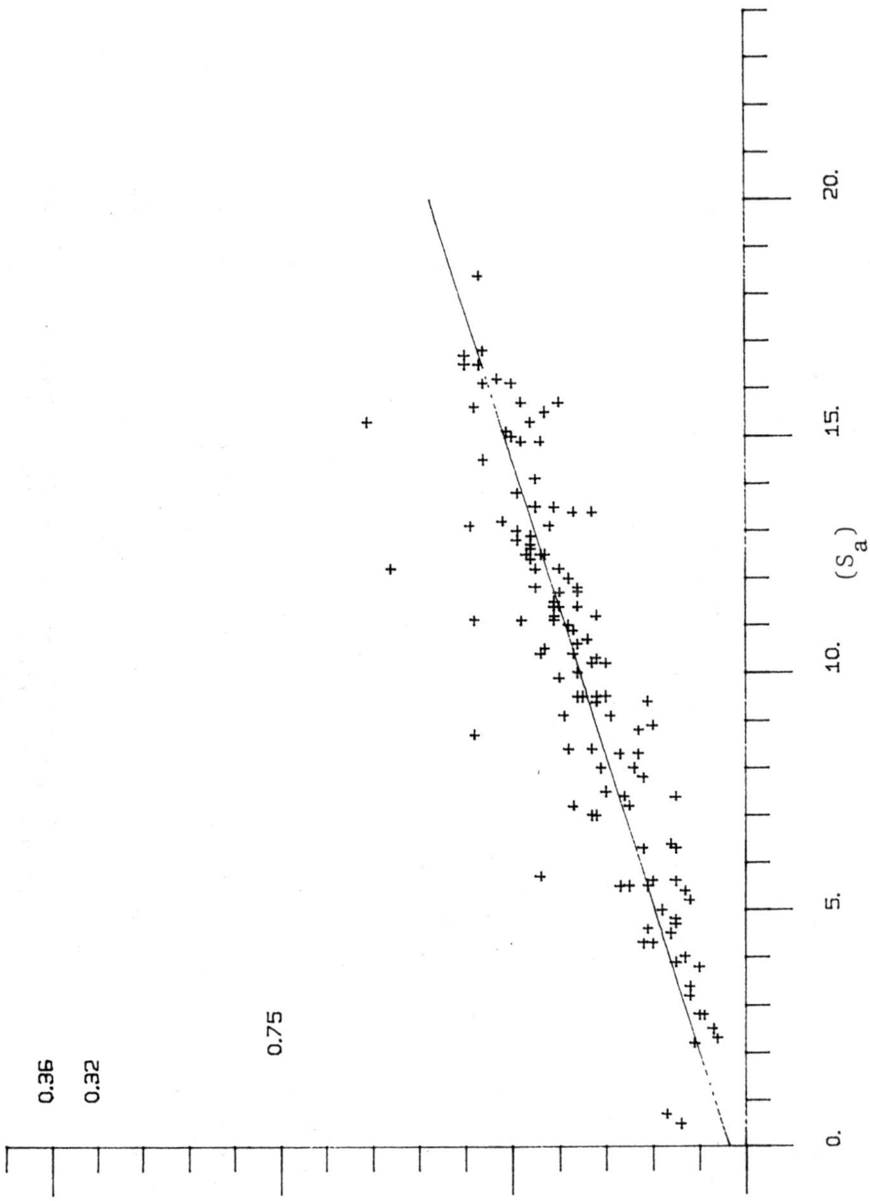

Fig. 8 Plot of ambient vs. living room swings for house #342 showing best linear fit. #342 is a heavyweight house.

Table 3

House	Mass	A	B	C
359	lighweight	0.84	0.84	0.79
357	brick veneer	0.77	0.68	0.74
350	conc. slab	0.57	0.44	0.71
351	conc. slab	0.63	0.44	0.76
346	conc. slab	0.67	0.28	0.84
345	slab, bv walls	0.45	0.33	0.66
340	slab, bv walls	0.59	0.30	0.66
344	cavity brick	1.00	0.35	0.39
349	cavity brick	0.92	0.25	0.46
342	slab, mass wall	0.36	0.32	0.75
347	slab, cavity	0.63	0.29	0.54
341	slab, cavity	1.06	0.25	0.52

Table 4

House number	LR v. BR A'	B'	R2'	LR v. Ambient R2
341	0.68	1.12	0.66	0.52
349	0.53	0.88	0.73	0.46
351	0.70	1.16	0.75	0.76
357	0.43	1.03	0.70	0.74

V. OUTLOOK

V A. EXTENSIONS OF THE ANALYSIS

The monitoring program at Bonnyrigg will continue for the next three years. Houses will remain unoccupied for the winter season 1982, and tenants will move in subsequently. Ultimately it is the performance of occupied buildings which is important. However, a knowledge of building characteristics is vital to the understanding of occupied behaviour. Accurate estimates of these building characteristics can only be obtained under controlled, unoccupied conditions.

Additional data which we expect to acquire during the winter campaign of 1982 will enable us to complete the analysis begun here. It should be possible to find a seasonal dependence for the solar gain coefficient. A larger data set will allow us to find meaningful correlations between the shell heat loss coefficient and wind conditions. Wind direction, which we have not mentioned, is likely to assume an important role in the variation.

Electric heaters will be operated on thermostatic controls for thirteen of the fifteen houses during the winter. Daily power consumption figures will make it possible to re-evaluate the energy equation and solve for actual coefficients rather than ratios of coefficients as we have done here. This information will place the solar gain coefficient in proper perspective and allow an accurate assessment of insulation levels achieved.

The problem of relating internal fluctuations to external fluctuations in a statistically meaningful way may be resolved by the introduction of the concept of "no-mass swing". The no-mass swing, briefly, is the internal temperature swing in a building without any thermal capacity. Given a building under consideration, its no-mass counterpart is an abstraction with identical shell, floor, and radiation characteristics, but no thermal mass. Given the weather and ground temperature, internal daily temperature swings are easily calculated for such an abstract building. The the comparison of measured swings and no-mass swings will yield a mass function which takes realistic account of not only ambient air temperature and insolation, but ground temperature as well.

An alternative method of employing regression analysis to derive building characteristics is to operate in the frequency domain. By employing daily averages and disregarding sequential nature of information from different days we have chosen to operate in the time domain. The price we pay for this choice is a clumsy treatment of processes involving time lags, such as thermal storage. On the other hand if we acknowledge the sequential nature of the continuous stream of hourly measurements then the powerful tools of time-series analysis may be employed. However a longer period of continuous data acquisition is necessary if these tools are to be employed fruitfully.

V B. APPLICATIONS

What began as a presentation of empirical results became, out of a need to simplify the presentation, a worked example of a new analytical tool. Although it resembles other simplified prediction methods it has the weight of hard measurements behind it. Here, the building coefficients are under investigation, not the indoor environment. Simplified methods connote the neglect of certain processes in an attempt to achieve a first-order approximation. Certainly we have neglected some influences, but the search for overall effects is more characteristic of our approach than the search for principal effects. We have treated the building as a weather filter. We were interested in the characteristics of the filter and not the complex processes by which filtering is achieved.

In this sense, our approach was independent of any models regarding the filtering processes. Therefore these results may prove useful in the testing of simulation models. Additionally, the regression analysis may be applied to a very wide range of buildings, provided one may define an isothermal zone which is characterized by Ti and Si. The testing of models could proceed in two ways. Either coefficients corresponding to SGC, SHLC, FHLC, and the mass function would be mathematically derived from the model, or these coefficients would be derived from simulated data in the same fashion as we derived them from measured data. In either case, comparisons would be made on a coefficient-to-coefficient basis. Discrepancies would be manifestly clear. The model aspects giving rise to the discrepancies could be isolated.

Finally, it might prove possible to specify standard coefficients for standard construction types in the form of a look up table. With these numbers available a computationally efficient method using real weather data and employing only linear equations would be capable of producing daily averages and swings in internal temperatures to within much smaller error limits than simplified methods can currently achieve.

ACKNOWLEDGEMENTS

The authors wish to thank Terry Williamson for his suggestions regarding the mathematical approach. The Bonnyrigg Solar Village Project was funded by the National Energy Research Development and Demonstration Council, by the Energy Authority of New South Wales, and by the University of New South Wales. The project was sponsored by the Housing Commission of New South Wales.

REFERENCES

(1) ANSI standard C96.1

(2) Carnahan, Luther, Wilkes. Applied Numerical Methods. John Wiley and Sons, Inc. New York 1969.

(3) Jenkins, Watts. Spectral Analysis and Its Applications. Holden-Day. San Francisco 1968.

NOMENCLATURE

Cv	Heat capacity of building (W-hr/deg K)
SGC	Solar gain coefficient (m^2)
SHLC	Shell heat loss coefficient (W/ deg K)
FLHC	Floor heat loss coefficient (W/ deg K)
$\overline{Ti}$	Mean internal air temperature over one day (deg K)
$\overline{Ta}$	Mean ambient dry bulb temperature over one day (deg K)
$\overline{Tg}$	Mean ground temperature over one day (deg K)
$\overline{I}$	Mean global horizontal solar radiation over one day (W/m^2)
Si	Daily swing in internal air temperature (deg K)
Sa	Daily swing in ambient air temperature (deg K)
P1	The ratio SGC/(SHLC+FHLC) (m^2degK/W)
P2	The ratio SHLC/(SHLC+FHLC) (dimensionless)
P3	The ratio FHLC/(SHLC+FHLC) (dimensionless)
P4	The ratio Cv/24* (SHLC+FHLC) (dimensionless)
Qam	Instantaneous rate of heat absorption by mass (W)
Tm	Representative present temperature of thermal mass (deg K)
Tm'	Representative mass temperature 24 hours ago (deg K)
MF(*)	Mass function to be determined from data
Yk	Dependent variable in regression
Xik	Independent regression variables
$\hat{Y}k$	Estimate of Yk
σ	Square root of residual variance using least-squares estimates of coefficients. (deg K)
A	y - intercept of best-fit line
B	slope of best-fit line
R2	Correlation coefficient for LR v. Ambient regression analysis
R2'	Correlation coefficient for LR v. BR regression analysis

Index